The Helicopter Pilot's Handbook

Phil Croucher

About the Author

Phil Croucher holds EASA, UK, UAE and Canadian licences for aeroplanes and helicopters and around 8700 hours on 37 types, with a considerable operational background, and training experience from the computer industry. He is currently the Head Of Training for Caledonian Advanced Pilot Training, the European licence conversion specialists. He can be contacted at **www.electrocution.com**

Contents

1 Introduction   1-1

The Job   **1-1**
Getting Started   **1-2**
Going For A Job   **1-3**
  The Advert   *1-5*
  Your Resume   *1-5*
  The Interview   *1-7*
What Type Of Company?   **1-10**
  Scheduled   *1-11*
  Charter   *1-11*
  Corporate   *1-12*
Company Personalities   **1-13**
  The Managing Director (or CEO)   *1-13*
  The Chief Pilot   *1-13*
  Flight Safety Officer   *1-13*
  The Chief Training Captain   *1-14*
  Base Manager   *1-14*
  The Maintenance Contractor   *1-14*
  Engineers   *1-14*
  The Operations Manager   *1-17*
  Quality Assurance/Compliance Manager   *1-17*
  The Company Pilot   *1-17*
  The First Officer   *1-19*
  Others   *1-19*
  Customers   *1-19*
The Operations Manual   **1-20**
Flight Time & Duty Hours   **1-21**
  Your Responsibilities   *1-22*
  Company Responsibilities   *1-22*
  Maximum Duty Period (FDP)   *1-22*
  Discretion to Extend FDPs   *1-22*
  Minimum Rest Periods   *1-23*
  Discretion to Reduce a Rest Period   *1-23*
  Split Duties   *1-23*
  Positioning   *1-23*
Setting Up A Company   **1-24**
  Financial Matters   *1-24*
  Obtaining An AOC   *1-28*
  Running Things   *1-29*

2 Qualifications   **2-1**

Licences   **2-1**
   The CPL(H)   *2-1*
   ATPL(H)   *2-1*
EASA   **2-1**
   Flying Training   *2-2*
Canada   **2-2**
USA   **2-3**

3 Specialised Tasks   **3-1**

Aerial Application   **3-1**
   The Rotor Wake   *3-1*
   The Procedure   *3-2*
   Spray Drift   *3-4*
   Seeding   *3-4*
Fire Suppression   **3-5**
   Bucketing   *3-7*
   Aerial Ignition   *3-8*
EMS/Air Ambulance   **3-9**
Pleasure Flying   **3-11**
   Running The Site   *3-11*
   Standard Rescue Equipment   *3-11*
   Site Dimensions (for UK)   *3-13*
Special Events (for UK)   **3-15**
Line Patrol   **3-16**
   Wires   *3-17*
   Power Line Cleaning & Maintenance   *3-17*
Pipeline Survey   **3-18**
UK Police Operations   **3-18**
   Prisoners   *3-18*
   Persons under the Influence   *3-18*
   Police Dogs   *3-18*
   Weapons and Munitions   *3-19*
   Bodies and Remains   *3-19*
   Formation Flying   *3-19*
   Landing Helicopters on Roads   *3-19*
Hover Emplaning and Deplaning   **3-20**
Parachute Dropping   **3-20**
Aerial Filming and Photography   **3-21**
   The Movies   *3-21*
Aerial Survey   **3-22**
Air Testing   **3-22**
Seismic Support   **3-23**
   Flight Following   *3-24*
   The Township System   *3-24*
Avalanche Control   **3-25**
Aerial Harvesting   **3-26**
Wildlife Capture   **3-26**
Frost Control   **3-27**

Hazards		*3-27*
Preparations		*3-27*
Operational Procedures		*3-27*
Responsibilities		*3-28*

4 External Slung Loads — **4-1**

Admin	**4-2**
Ground Crews and Equipment	**4-2**
Ground Equipment	*4-3*
Static Electricity	*4-3*
Helicopter Condition	*4-3*
Sling Equipment	*4-4*
Hooks	*4-4*
Lines	*4-6*
Loading and Unloading Areas	*4-11*
Performance	*4-11*
Preparation of Loads	*4-12*
Personnel Briefing	*4-13*
Hooking Up	**4-14**
Load Behaviour	**4-14**
Vibration	*4-15*
Swinging	*4-15*
Setting Down	**4-16**
Vertical Reference (Longlining)	**4-17**
Depth Perception	*4-18*
Getting Started	*4-19*
Getting It Back On The Ground	*4-22*
Tips &Tricks	*4-23*
Typical Hook Loads at MAUW HOGE	**4-24**

5 Mountain Flying — **5-1**

Safety	**5-1**
Performance	**5-2**
Pressure Altitude	*5-2*
Density Altitude	*5-2*
Illusions	**5-4**
Air Movement	**5-5**
The Demarcation Line	*5-6*
Winds	*5-8*
Turbulence	*5-10*
Downdraughts, etc	*5-10*
Mountain Waves	*5-11*
Valley Flying	**5-15**
Lee Slopes	*5-16*
Landing Sites	**5-17**
Log Pads and Platforms	*5-18*
Approaches	**5-18**
Finding The Wind	*5-20*
The Eye Level Pass	*5-21*

Maintaining A Sight Picture		*5-22*
Peaks		*5-22*
Cirques (Bowls)		*5-23*
Canyons		*5-24*
Narrow or Dead-End Valleys		*5-25*
Riverbeds		*5-26*
Cols (Saddles)		*5-26*
Takeoffs		**5-27**
Downwind		*5-27*
Heli Skiing		**5-28**
Confined Areas		**5-28**
6	**Search & Rescue/Hoisting**	**6-1**
Definitions		**6-2**
Levels of SAR Service		**6-2**
Level 1		*6-2*
Level 2		*6-2*
Level 3		*6-2*
Planning		**6-3**
Weather & Other Conditions		**6-3**
Equipment		**6-3**
Winch Cable		*6-3*
Utility Hoist		*6-3*
Rescue Sling (Horse Collar)		*6-3*
Hoist Harness (Double Lift Harness)		*6-3*
Rescue Net (Billy Pugh)		*6-4*
Stokes Litter		*6-4*
Additional Miscellaneous Gear		*6-4*
Search Patterns		**6-4**
ELT Aural Homing		*6-4*
Expanding Square		*6-4*
Track Crawl		*6-5*
Parallel Search		*6-5*
Sector Search		*6-5*
Coast Crawl		*6-5*
Contour Search		*6-5*
Communication & Protocols		**6-6**
Flight Following		*6-6*
The Commentary		*6-6*
Line And Hover Corrections		*6-6*
Speed Corrections		*6-6*
Height Corrections		*6-7*
Standard Terminology		*6-7*
Emergency Terminology		*6-8*
Standard Circuit		**6-9**
Stage 1		*6-9*
Stage 2		*6-9*
Stage 3		*6-9*
Stage 4		*6-9*
Stage 5		*6-9*

© Phil Croucher, 2016

 Actions & Commentary   *6-9*
The Hoist   **6-11**
 Single Lift   *6-11*
 Double Lift   *6-11*
 Double Strop Hypothermic Lift   *6-11*
 Stretcher Transfers (SAR)   *6-11*
 Downwind Checks   *6-12*
 Final Approach   *6-12*
 Who Has The Con?   *6-12*
 Return To The Rest Position   *6-12*
 After Hoist Checks   *6-12*
Emergencies   **6-12**
 Engine Failure   *6-12*
 Loss Of Reference   *6-12*
 Hoist Freeze   *6-13*
 Cable Runaway In/Out   *6-13*
 Ditching   *6-13*
 Load Swing/Spin   *6-13*
 Intercom Failure   *6-13*
 Fouled Hook   *6-14*
Safety Matters   **6-14**
 Restraint Harness   *6-14*
 Safety Lanyard   *6-14*
 Cable and Gloves   *6-14*
 Hoist Hook   *6-14*
 Delivery   *6-15*
 Retrieval   *6-15*
 Securing   *6-15*
 Slack Cable   *6-15*
Hoisting From A Vessel   **6-15**
 Approaching The Vessel   *6-15*
 Outbound   *6-15*
 Inbound And Final   *6-15*
 On Top   *6-15*
 Lowering The Rescue Device   *6-16*
A Real Rescue   **6-17**

7 Offshore   7-1
Helidecks   **7-1**
Personnel   **7-2**
Procedures   **7-2**
 Takeoff & Landing   *7-3*
Ship's Deck Landings   **7-6**
 Landing Areas   *7-6*
 Ship Motion   *7-6*
 Communications   *7-7*
 Approach & Landing   *7-7*
 Departure   *7-7*

8   Operational Stuff   8-1

VFR En-Route Minima   **8-1**
    Over Water   *8-1*
Landing Sites   **8-1**
Fuel   **8-2**
    Fuel Management   *8-4*
    Drums   *8-5*
    Fuel Checking   *8-6*
    Passengers on Board   *8-7*
Ditching   **8-7**
    Sea Movement   *8-8*
    The Procedure   *8-8*
    Equipment   *8-9*
Remote Areas   **8-10**
    Ground-Air Visual Signals   *8-11*
    Air-Ground Visual Signals   *8-11*
    Droppable Containers & Packages   *8-11*
Emergency Equipment   **8-11**
Icing   **8-12**
    Ground De-icing   *8-12*
    General Precautions   *8-12*
Recording Of Flight Times   **8-13**
Passenger Safety   **8-13**
Night Flying   **8-14**
    Schermuly Flares   *8-15*
Winter Operations   **8-15**
    The Weather   *8-15*
    Snow   *8-18*
    Whiteout   *8-18*
    Taking off   *8-19*
    The Cruise   *8-19*
    Navigation   *8-20*
    Landing   *8-20*

9   Techie Stuff   9-1

Twins   **9-1**
Performance   **9-1**
    Definitions & Abbreviations   *9-2*
    V-Speeds   *9-5*
    Airworthiness Categories   *9-6*
    Performance Classes   *9-8*
    Class 1   *9-8*
    Backup/Lateral Procedures   *9-9*
    Class 2   *9-11*
    Class 3   *9-14*
    Procedures   *9-14*

INTRODUCTION

O ne snag with helicoptering is that there are virtually no flying clubs, at least of the sort that exist for aeroplanes, so pilots get very little chance to swap stories, unless they meet in a muddy field somewhere, waiting for their passengers. As a result, the same mistakes are being made and the same lessons learnt separately instead of being shared. Even when you do get into a company, there are still a couple of things they don't teach you, namely that aviation runs on paperwork, and how to get a job, including interview techniques, etc. Flying the aircraft is actually a very small part of the process, as there are many other factors that influence your working life (see *The Job*).

One significant drawback is that nobody really *tells* you anything, either about the job you have to do (from the customer) or how to do it (the company) - you will always be up against the biggest hazard to flight safety - the other guy who did it last week! Sure, there will be training, but, even in the best companies, this can be minimal, and definitely not standard around the industry. Also, what you do for the rest of your career is very dependent on the first company you work for.

This book is an attempt to correct the above problems by gathering together as much information as possible for helicopter pilots, old and new, professional and otherwise, in an attempt to explain the *why*, so the *how* becomes easier (you will be so much more useful if you know what the customer is trying to achieve). In short, it is the result of all the mistakes I made because there wasn't a book like this to refer to, containing the stuff I had to pick up on the way which, hopefully, you won't have to!

Although most of its contents are based on my own experience, the gaps owe a lot to many people who have either taught me a few things along the way or contributed material, including Doug Potts, Tony Boulter, Peter Boitel-Gill, John Wood, John Marsden, the late great John "Chalky" White, John Woodhouse, Neil Laird, Ron Howard, Dave Moss, Dave Ankin, Ray Portlock, Paul Smith, Shawn Coyle, Don Plattel, Jim Filippone, Jeff Mitchell, Dan Beck, Terry Clancy, Paul Johnson, Northern Mountain Helicopters, Remote Helicopters, CHC, Aerogulf and several others, especially Jim Gray, Peter Rover and other Transport Canada or CAA members, many of whom have asked to remain anonymous.

Special thanks to Paul Gibson for his LongRanger info and Don (407Driver) for the longline picture!

Otherwise, all I have really done is gathered up all those bits of paper lying around various training offices and crewrooms and assembled them in some sort of order. I don't know the authors, but I'm sure they would be glad to know that the fruits of their experience are gaining a wider audience. If anyone recognises anything, please get in touch and I will be glad to mention you here (in this respect, extra special thanks to Brad Vardy!)

Note: *This is a book for professional pilots.* As well as using standard aviation phrases as shorthand, it is assumed that activities will be carried out safely, with any equipment required being approved and serviceable. Also, all operating limits will be observed and remaining engines (when one fails) will operate within maximum continuous power conditions.

THE JOB

Flying helicopters is definitely not a 9-5 affair, and it's the only work some people ever see themselves doing. You are certainly part of a select brotherhood, but there are things you need to know before doing it professionally, which is a different ball game than flying for pleasure - the rules are entirely different. Mostly, the "extras", in the shape of paperwork, responsibility and office politics, are bearable for the amount of pay you get, and the flying itself is the icing on the cake.

It can be tiring, for a start. Four hours' worth has been compared to eight hours' hard labour and double that when longlining, all due to the concentration, particularly in mountainous areas and going in and out of clearings all day long (or over 120 landings a day with heli-skiing), where you cannot afford to let your attention slip. Do this for 21 days on the trot (sometimes up to 60) and you will also find you need to keep physically fit in order to cope.

However, don't expect to be flying all day and every day. There will be a lot of waiting around in the back of your machine in remote places, so you will need plenty of books (that's how I wrote this!) A typical day on duty may be up to 10 or 12 hours long, within which you might fly for about 3 or 4. The gap between them will be larger in the Corporate world, but not always.

When I made the move to RW, I thought I knew quite a bit about hands and feet type flying, wind, mountain work, high altitudes etc., etc. - I was in for a surprise! Those environments are very tough in a light airplane, but are much more complex in a helicopter, with the consequences of a screw-up coming faster and harder.

I've also been very impressed with the maintenance and overall attitude in the Rotary world. Things are done better, AME's are more concerned with getting things done right than quickly, and the level of pay is much better. I think this probably has something to do with it.

Perhaps the most interesting difference is the attitude. In Utility rotary work, you need the skills to do the job, long-lining (seismic, drill moves, construction), mountain work etc. If you don't have those skills, you simply cannot do those jobs. In IFR FW, much like IFR RW, you don't need these hands and feet/decision making skills, and most anybody can be put in the seat and do a reasonable job of it. Because of that, there are HUGE egos walking around, thinking they're God's Gift, but in reality, their greatest ability lies in tieing a half-windsor and applying hair gel in the morning.... In Utility RW work, you walk the walk, or you just don't talk. You can move that diamond drill, or you can't, and everyone knows it. In FW, most pilots I came across in the IFR side of things couldn't fly their way out of a paper bag, yet would have you believe they're the second coming of Geoffrey DeHaviland/Chuck Yeager depending on your country of origin.

I think there is a level of pride in VFR rotary work that doesn't exist in most FW/IFR RW applications, because of the level of skill involved. Now, RW pilots are not the spit and polished bunch that most FW guys/girls are, but their professionalism lies in the way they do their jobs, not how they're dressed, or the Mach number their machine cruises at.

And again:

Flying fixed wing is generally where you have not a lot to do and someone to help you do it. Rotary wing is where you have it all to do and NO-ONE to help you do it. The fixed wing flying job usually ends at the destination. The rotary wing job usually BEGINS at the destination.

Anybody can land, then stop - it takes a special kind of person to stop, then land!

You also have to remember that helicopters are generally used because there is no other suitable means of transport, which means going to places nobody's ever heard of, in strange weather, and not being home for weeks on end. In Canada, this is very much part of the job, so think twice if you don't think you can cope (as far as family life is concerned, flying helicopters is similar to running away and joining the circus!) If you want to get home at night, have a look at corporate, offshore or instructing.

Finally, some of it involves making money out of other peoples' misfortunes, as when reporting on disasters, etc., which is an aspect I have never really liked, but helicopters save lives, too and, as I said, it's the only thing some people want to do, whether for the flying or the lifestyle, so let's have a look at how you become a helicopter pilot.

GETTING STARTED

This is the most difficult bit - the cost of helicopter time is so great that it's almost impossible to do without help, maybe from parents, or being trained in the Forces. Having said that, there are plenty of people who have done it, so it isn't impossible, but these will tend be found in Canada or the USA, where it's considerably cheaper. In Europe, where it's over twice as expensive and you need more hours to get your licence, anyone who can afford their own training would, in terms of financial reward, have to think twice before working as a pilot, because that sort of money is considerably more productive elsewhere. At least then you can fly when you want to.

Mind you, it's ultimately not that different in North America. Even though you only need 100 (Canada) or 150 (USA) hours to get your ticket, you are still usually unemployable, unless your family owns the company (and even then the insurance companies or customers would have something to say), so you either have to do a couple of years as a hangar rat, that is, washing windscreens until your company sees what you're like and trains you up, or buy the hours yourself. To be even remotely interesting to an employer (or, more correctly, a customer), you need at least 500 hours, sometimes 1000 or 1500, or some sort of specialised training, such as a mountain course (or both) and maybe an instrument rating, depending on the job. In the US there is almost a set regime of being a student from 0-200 hours, an instructor or SIC from 200-1000 hours, helicopter tours or oil rig work from 1000-2500 and whatever you choose after that.

Larger companies may run courses for ground staff with commercial licences, and who have been observed for a couple of seasons for "suitability". It will be run by senior pilots who are also instructors, and is a good thing to get on, as it will improve your prospects over other pilots with the same hours as you, assuming that other companies recognise the standards. In fact, your training background is so important (when there's nothing else to go by) that you should pick your school carefully if you can't get on such a course. Make sure whoever teaches you has actually been out and done the job themselves, and have maybe run their own companies. Unfortunately, people can become instructors at 200 hours and stay there. Granted, some of them are real good at their jobs, and if this wasn't possible, the industry wouldn't have nearly as many pilots as it needs, but there are many who would prefer that instructors have at least 1000 hours before they start, because trouble is best avoided by not getting into it in the first place, and you only know how to do that with experience. You can't teach what you don't know.

A lot of schools indicate that they might hire you once you complete your training, but don't include that as a factor in your choice, as it's generally only those that are part of a larger commercial organisation that can afford to do it, and the competition is keen. The employment situation can change from day to day, and it can be impossible to keep up with. Just regard it as a bonus (indenturing employees for "training" was quashed by the Supreme Court of Canada in 1968, so ignore those "contracts").

When you budget for your training, don't just count in the cost of your course, but the time afterwards going around companies to get hired; just sending resumes is no good at all (see *Going For A Job*). You may also need more hours than you think - the average time taken to pass the PPL is 67.7 hours, against a minimum requirement of about 40. The ratio is not much different for commercial licences.

The next chapters look at what you need to do to get your licence, then what you might get up to after that, so you know what you're letting yourself in for (if it's any consolation, lawyers and photographers get the same treatment). First of all, though, you need to get a Class 1 medical, because all the training in the world will useless if you fail it. Then do a trial lesson in a helicopter, to see if you actually like helicopters (or, more to the point, whether they like you!), then maybe do a PPL, so if you get the chance for some positioning flights, or you know someone who owns a helicopter, you can do them and build some hours (on the average commercial course, you don't get any sort of paperwork until the end).

I would worry about the commercial licence itself much later on in the process.

Otherwise, there are distinct stages in the average pilot's career. First, you fly single-engined piston machines, then turbine ones, then multis, then you might go IFR (in fixed wing, you go multi, IFR, then turbine). Along the way, you pick up specialist stuff like longlining, and by the time you retire you finally have enough qualifications to get a job! Employers want people with 10,000 hours, type ratings on everything made since the Wright Brothers, a Master's degree, the willingness to carry baggage and clean helicopters if they're not already helping the engineer maintain them, have worked for one outfit only during their career and left only when the owner died, and willing to work everywhere on a moment's notice for undefined periods, plus 2000 hours single pilot IFR, 2000 hours precision and/or production longlining, with 2000 hours of that above 7000 feet, plus 2000 hours offshore. And be younger and less qualified than the Chief Pilot!

GOING FOR A JOB

Inexperienced pilots have a similar problem to people in many other walks of life - they cannot get a job because they don't have the experience, and they cannot get the experience without a job (try being a junior lawyer). When looking for work with hardly any hours and a licence which is barely dry, you are in a similar position to asking your father for the keys to his new Mercedes so you can go to a party. You have to ask yourself what characteristics you might have that would make your father do such a stupid thing. Or that might make anyone get into it with you at the controls, for that matter.

What would your father want to know? That he will get his car back, of course, undamaged, and with no after effects, like traffic tickets. Similarly, a Chief Pilot will need reassurance that you are capable of flying one of the company machines without crashing it, upsetting the customers and being the cause of a subsequent visit from your local friendly Operations Inspector. In this respect your flying ability counts for only a small part of the qualities required - it's the remainder that need to be emphasised when doing the rounds at such a disadvantage (even failing your exams proves persistence if you finally pass them!)

OK, so now you're a Chief Pilot - what would you like to see in someone who walks into the office with a resume in one hand and no doughnuts in the other?

I would suggest a selection from the following would be appropriate:

- A smile on your face.
- A firm handshake.
- Confidence.
- Presentable appearance, including clothing and hairstyle - no shaven heads or curly locks, and especially no earrings.
- Clean vehicle.

It's a fact that jobs have been offered just on appearance. I know, because it happened to me, and no-one even asked to look at my logbook or licences (actually, the reason was because my resume was printed, in the days when it was a major achievement to get one typed). However, in the normal course of events, for low-timers, visiting as many companies as possible is about the only way to get yourself known. Just sending a resume is not good enough when they haven't seen you before.

Believe it or not, someone with relatively low experience and who gets on with customers is actually in a better position than somebody the other way round, other things being equal, as experience and flying techniques can be taught - personality can't. Also, get to know lots of people at the bottom levels, because Chief Pilots very often ask the guys on the shop floor if they know anyone when there's a vacancy and, if you are recommended, there's less chance of personality conflicts later (Chief Pilots don't like hassle, but they do like people who are not going to drop them in it, as they carry a lot of responsibility). At least one company I know of gets its pilots in the crew room whenever someone is about to be offered a job, and they take a vote.

Employers like people who have clearly made an effort to know their (potential) jobs, and who clearly absorb information and knowledge about their aircraft and other crew members. In other words, the sort of people who give every impression of being commanders in their own right and can be relied upon in flight and otherwise. Although much of this comes from experience, the potential is often very obvious at an early stage.

Remember also that loyalty goes both ways. Some companies deserve all they get when their pilots disappear in a shortage - with no staff, they can't trade, and they go out of business. It's happened before and will happen again (they forget that companies need good people, but good people don't need companies). On that basis, if you're doing the traditional two years as a hangar rat

before you get your hands on a machine, be prepared to move on if it seems like the company are more interested in your cheap labour than training you. In my opinion, in with your normal windscreen-washing, you should be doing the air tests and non-revenue flying, which will not only give you an incentive, but make your subsequent training cheaper by keeping you current. It is entirely possible to get well upwards of 400 free hours a year in a busy company, if you're prepared to end up in strange places for days at a time.

You will have to do a bit of research about every company you target - you will certainly need the name of whoever does the hiring, and the head of the department you want, if they are different (in most cases, it will be the Chief Pilot or Base Manager, or, in other words, someone with local knowledge). Only go to the personnel department as a last resort, and even then just to ask for the right name(s). You need to know the sort of work they do, the type of customers they have, where they operate, and tailor your initial conversation around it, emphasising the benefits you can bring which cause them the least amount of work. For example, in Canada, one of the first questions you will be asked is if your PPC (*Proficiency Check*) is current, because it can be transferable between companies if they operate the same machinery, and they won't have to spend money sorting you out. It's almost guaranteed that the next question will concern either a mountain course or long-lining experience, so be prepared. The point is that their requirement for a pilot is to solve a problem, and you need to be the one with the solution, so get their attention, then create the desire to employ you and, more importantly, do something about it. In fact, the sort of telephone conversation a busy Chief Pilot up to the ears in paperwork would like to hear is something like:

> *"Hi, I'm an Astar pilot with 1500 hours, mountain and longlining time, available now."*

Music to the ears. Just adjust it for your own situation, but only get detailed after you start fishing for what they want. If you get asked any question at all, you've got what is known as a "buying signal", but the question will likely come after a short period of silence, which you shouldn't break. Answering apparent brushoffs with further questions should keep the conversation going. If you can introduce the name of somebody already known to them, so much the better.

The Advert

If there is one, it's usually the last resort for companies who need staff - apart from being outdated anyway, the best jobs are almost always filled by word of mouth, and the ad is placed to satisfy legal requirements, or to wind up the opposition. In fact, the way an advert is worded can tell you much about the company you may be working for.

Read what it actually says. If it states definitely something like "must have 500 hours slinging", it means your application will go straight into File 13 (the waste bin) if you don't. On the other hand, another might say that such experience "is desirable" or "is an advantage"; if you score 6 out of 8 on the requirements, then go ahead. In this case, circumstances will determine what happens to your application, for instance whether there is a pilot shortage or not, or whether the Chief Pilot or the Personnel Department actually wrote the advert (Personnel won't haven't a clue as to what's really required and may have just copied it from somewhere else). Just bear in mind that words like "preferable" also become criteria for *weeding out* applicants if there are a lot of them.

However, your face may fit better than more highly qualified people, and it's a favourite hobby of some pilots to keep applying for jobs anyway, so to help you get on where you may be at some sort of disadvantage (whether you're one of many applicants or you haven't quite got the qualifications required), you may need to employ a few tactics, including your resume.

Tip: One tactic that sometimes works is to apply relatively late, say a week after the ad appears, ensuring that the bulk of applications are out of the way and whoever has become cross-eyed looking at them will get yours when he's back to normal, possibly all by itself so you're noticed more. You also (theoretically) go to the top of the pile. However, *do not miss the deadline* as, even if the Chief Pilot wants you, Personnel will bounce you out anyway. Another is to always make a follow-up call, including after an interview - in some companies, the process is very long and you can easily get forgotten.

Tip: If the ad runs again in a very short time, it means they haven't found anybody - if you didn't have the qualifications the first time round, you may do now!

Your Resume

Applying for a job involves selling yourself, by which I mean that you are the product to be marketed, and the process starts even with the envelope in which you send your details (a full-sized stiff-backed one ensures they don't get creased). It's surprising how many people fail to use the resume and covering letter (they are, after all, a first introduction) as properly as they should be. I have seen very badly handwritten resumes with no idea of spacing on ragged paper that would disgrace a fish and chip shop. This type of introduction says little for your self-image and is likely to go straight into the bin - if it doesn't, it will be a reminder of what you were like long after the interview. *Your resume is your sales brochure.*

Having said all that, in a lot of aviation companies the atmosphere is relatively informal, and, although you need a resume, hardly anyone ever reads it, at least not till you make them do so by turning up on their doorstep, so take the following remarks with as large a pinch of salt as you feel able. You may only be required to fill in an application form (see below), which will also involve a breakdown of hours - usually First Pilot and Grand Totals. The initial contact could well be a faxed one-page letter, with everything relevant on it, and full details when asked.

Tip: Keep a running breakdown of your hours, separate from your logbook and updated monthly, say, in a spreadsheet, which will help you extract these figures when required (it will also be a back-up should the original get lost, but a logbook must fulfil certain legal requirements). Keep columns for specialised stuff.

However, a large company with a personnel department (which therefore deals with several other professions) will expect to get the full treatment. Like flying, the more preparation that goes into your resume, the better the results you will get. Remember, you're trying to beat the opposition, in an environment where the best person for the job frequently gets eliminated early on, and the person who plays the application/interview game best wins. Unfair? Yes, but life's like that, so here's a couple of points to note before we go any further - the resume is not meant to get you a job, but an interview. Secondly, it actually consists of two parts - the resume itself, which contains the usual stuff, and a covering letter, which, being a business document, should be neatly typed or wordprocessed on white letter-sized paper, unless you are specifically told to do otherwise (you might be asked to fill in a form) - it looks more professional anyway.

The letter is actually a focussing device that should include information that might not belong in the resume, or to highlight anything that might be particularly relevant (from the ad, maybe) and to get it in front of the right person. Ring up to make sure you spell their name right, as "Dear Sir" or "Dear Madam" will often mean termination with extreme prejudice immediately. You may also include reasons for wanting to join the company, or, more to the point (salesmanship again), how useful you will be to them, because that's what they're bothered about. You could, for example, cover points mentioned in the advert, or you know that they're concerned about. This is your sales pitch.

Use the word "I" as little as possible, include any reference numbers in the advert, and get the person's job title right. Don't "wish" or "hope" for an interview (salesmen are taught to ask for a sale, so - ask for an interview!). Remember that most resumes look the same, especially if you use a Microsoft Word template!

If you are not replying to an ad, remember that Personnel often do not know about vacancies until they are actually asked to do something about one, so you need to get hold of the person in charge of the department or base you are interested in. One tactic might be to write to the Big Boss, whereupon it might filter down to the relevant person from above, giving them more of an incentive to do something about it. Don't be shy about this - speculative letters show initiative, which is one quality required when operating in remote places. It also saves them money, if they are actually looking, as recruitment costs money (when talking about yourself, and therefore saying nice things, when you begin to feel slightly embarrassed is the time to stop).

Although it is often said that a resume should fit on one page (and this is good advice), life is never so convenient, and you should always be aware from the start that you might need 2 or even 3, if you include a breakdown of your flying hours. On the one hand, trying to cut everything down when it won't get any smaller is stressful, and on the other, many resume readers (myself included) find it frustrating that more information isn't forthcoming when I want to read it. The trick is to put the information you think might be needed on the first page, and expand it on the following pages, even if you repeat yourself (you could also put it in the covering letter). As a guide, my own procedure is to go through any list of resumes with the requirements of the job in mind, and either highlight any that are already mentioned, or write down any that are not, on the front page as an aid to later sorting. What is relevant depends on the job, but it's a fair bet that licences, types flown, total hours on each and availablility would be a good start - you could probably think of more, but especially include contact details.

Don't bother with referees, as these are usually taken up after the interview anyway.

Having said all that, you should still try to get the information in as short a space as you can without leaving anything out - if you're only going for a flying job, the tendency to include irrelevant information should be avoided, and everyone knows what a pilot does, so your resume will be on the technical side, that is, short, competent and to the point. Management qualifications (if you have them) are not important to somebody who just wants a line pilot (all the advice here should be read in this light - you don't have to include everything).

As with all salesmanship, you're trying to make it as easy as possible for the customer, in this case your potential employer, or at least the poor clerk in the personnel office who has to go through all the paperwork before the interviews (it's worth mentioning at this point that the clerk's job is to screen you out, or to discover who *not* to interview). If you feel the need to be more specific, use the covering letter to get your details in front of the right person. The screening out can take place in as little as 8 seconds - the irony is that they use the resume for the process. What do they see in that time? Well, the type of paper, its condition and layout, to mention but a few items (your subconscious can pick up a lot without you knowing). In short, whether you've spent time on it.

You need to use quality paper, A4-sized and white, and therefore inoffensive, but this requirement is really for scanning. Use one side of the paper only with the script centralised, with no underlining or strange typefaces. Leave at least a one-inch border at the top and bottom of the page with a good sized margin on either side. It will cost a minimal amount to get a two-page resume wordprocessed properly and not much more to get a reasonable number printed, preferably on to the same paper. Use a spellchecker. Twice.

It should include your career history, commencing with your present position and working back about 5 years in detail, the remainder in brief. The name and town is enough to identify employers with a brief description of their activities, if needed, as aviation is a small world. You may include reasons for leaving your current position but, as said above, when people read a resume they almost always do it with a highlighter in one hand to mark relevant passages for later, and you can almost guarantee that this will be a prime target, so prepare it very carefully.

In summary, the layout must be neat, as short as possible, well spaced and easy to read, with a positive attitude conveyed throughout. If you don't have much experience, include outside interests that have transferable skills. All other personal stuff (date of birth, etc.) should be at the end, as it bores most readers. Here's a suggestion from someone who reads a lot of resumes:

Name

Address

phone, email

Personal Information

- 1 line with nationality and date of birth
- 1 line with education (highest schooling certificate)
- 1 line languages spoken (fluent/basic)
- 1 line with marital status/children

Flight Crew Licences (country)

- 1 line stating CPL, IR, ME...
- Other qualifications/licences (Flight Instructor)
- 1 line Medical Class/Date expires

Flight Experience

- Total hours XXXX
- Multiengine (less than 5700 kg)* - XXX hours PIC
- Single Engine* - XXX hours PIC
- Instrument - XXX hours
- Turbojet (and type aircraft) - XXX hours PIC/SIC

Current Passport

Availability: Immediate

*no need to list types if applying for an airline or large company.

The licence level will tell people a lot about your qualifications as pilots go through certain stages in their career. It shouldn't matter where you got your licence.

APPLICATION FORMS

Practice on a photocopy first, and always use the same pen throughout (that is, make sure you're not likely to run out of ink halfway through and have to change colours).

Don't leave blank boxes - use N/A (*Not Applicable*) if one doesn't apply, and never refer someone to an attached resume (that is, attach one if you like, but don't ask them to look somewhere else for information they want *now*).

The "other information" box is the same as a covering letter, so don't miss it out.

The Interview

Let us first of all establish what the interview is not. It has nothing to do with your competence as a professional, except for the simulator ride (if one is required). The fact that you've been put on any list at all, let alone shortlisted, indicates that your flying abilities are recognised.

On their side, the interview is really to see if your face will fit. They are about to let your personality loose on their customers and they want to see if you will help solve the problem or become part of it. In other words, they are looking for people who know the rules, have common sense, and the personality and tact to apply them. In other words, they are interviewing future *Captains*.

You, as an employee, must create value beyond the cost of employing you. As far as you are concerned, it's a chance to see if you will like the Company, in which case you may find it useful to write down what you want from them.

Note: With reference to value, mentioned above, the cost of employing you is not just your wages - you may have training or health insurance thrown in, plus other benefits, not to mention the staff employed to look after you, or any office you might have. In the first year, you may well cost much more than your salary. Even the interview process can cost thousands!

Interviewing techniques can be very sophisticated. You may be lucky and get away with a quick half-hour with someone who is just as nervous as you are, but the full-blown two day affair with Personality and Psychometric testing is becoming increasingly common. Certainly, it is used by one Electricity Board in the UK, and almost every airline worldwide. The full nine yards might include written maths, intelligence and psychological tests (with over 600 questions), a simulator ride, an interview and a medical (nine yards, by the way, or 27 feet, was the length of an ammunition belt in a B-17, so I'm told). Most questions are relatively simple, but the average time for each is about one minute. Examples are figuring out the next number or symbol in a logical series, identifying how a shape or object would look if rotated, finding words amongst a group of mixed letters, etc. The psychological part is not timed and presents situations and statements to be ranked from 1 to 4 according to which is the most or least like you. Do not try to read into questions or guess what they are trying to achieve, just answer them (don't add your own selections!) There may also be a team exercise, perhaps an evacuation plan for a village about to be flooded, in which you are given priorities and resources. There won't be a right answer - they will be looking for group interactions, such as who takes charge, who sits back and contributes quality input at the right moment, etc.

Anyhow, whatever shape it takes, you must regard the interview as having started whenever you walk through the main door of the building or meet any Company person. You are definitely under observation at lunch (why do you think so many people join you?), and the receptionist has been on the team more than once.

Tip: The problem with lunch (for you, anyway) is that it's an opportunity for many questions that cannot be asked elsewhere, so be even more on your guard.

The interview is therefore even more part of your sales technique. Naturally, you will be smartly dressed and presentable, and you must convince them that they are not so much buying a pilot as peace of mind. Unfortunately, most interviewers make their decisions about you in a very short time, based on what they see, feel and hear, well before the dreaded interview questions even start!

Although unlikely in a pure Aviation company, there may be questions or situations designed to put you on the spot by trying to destroy your composure. To combat this, there are ways of behaving that will give you the most confidence. Don't talk too much, don't be pushy or negative and don't break silences. Awkward questions are mostly to establish the pecking order should you actually join the company later; the answers, to them, are not that important. They may even be there to see how you handle stress and whether you can be intimidated (by passengers, maybe), and the only weapon you have is to practice beforehand, though it's best to pre-handle certain *types* of question rather than specific ones. You might, for example, be asked how your life will change if you are successful, or even whether you would be happier elsewhere. Most questions don't have a right answer - the interviewers are looking at your ability to reason, and justify your answers. Here are some possibilities, based on my own experience and that of others:

- Technical questions on the aircraft flown or from the pilot's exams, such as What is Dutch Roll?, or What are swept wings for?

- Tell us about yourself.

  You couldn't get much more open-ended than this. It is typically used as a warm-up, so keep away from your life history. Try a quick two or three minute snapshot of who you are and why you are the best candidate for the job.

- What if you smelt alcohol on your Captain's breath?

  Here, they are looking to see if you recognise a dangerous situation, know your responsibilities and have the tact to face up to a Captain. In the real world, in large organisations at least, much of your life will involve covering your backside against other peoples' mistakes or problems. As a first officer, it's your job to challenge the Captain if there is something wrong, but you don't want to give the impression that you're going to cause your own problems later. Reporting someone is almost always the wrong answer, at least for a first response, if only because it's bad CRM! Start with determining whether the aircraft is likely to be damaged (people don't usually deliberately screw up) and from that whether to terminate the sortie or not. In this situation, I would discuss it directly with the person concerned (making sure my facts were right) and leave it up to them to remove themselves from the flight (known as "passing the buck"). I would probably only take it higher if the Captain didn't take the opportunity to fall on his own sword or started intimidating me (escalating the problem is not only for your sake, but the people coming after you who might not have so much courage). In any case, I would ensure the discussion was in the cockpit (where the CVR would be running) or at least in front of credible witnesses.

- What would you do if you left your licence behind?

  You shouldn't take off without it - an insurance company could use it as an excuse not to pay up if an accident happened. Not only that, the company's operating certificate would be at risk. However, in the US, you could get an official copy faxed to you, which would be legal for 60 days.

- Why do you want to be a pilot?

  The view's good!

- What are your strong or weak points?

  One school of thought says that you should take one of your strengths and present it as a weakness, such as working too much, but I take the view that your time management skills may be lacking.

- How would you handle a grumpy captain?

  Gently!

- Why do you want to work for us?

- What is the biggest economic threat to this company?

- What does this company fly and to where?

- What is the Boss's name?

- What is the share price?

- Who did you train with?

- Tell us something that taught you about flying.

- Where do you want to be in 5 or 10 years' time?

- Have you ever scared yourself?

- Once the gear is up, what conversation will you have with the Captain?

  None that does not directly concern the operation of the aircraft.

- Put these in order of preference: Small children, Guns or Flowers

  Flowers (cargo makes more money), children do not weigh as much as adults, and guns require too much paperwork

- How would you deal with a personality clash with an arrogant Captain?

- You have the Board in the back, and they insist on getting to a meeting on time. 30 miles out, the destination is under a LVL 6 cell with very little movement. What would you do?

- Besides good pay (*yeah, right* - author) why do you want to be a pilot?

- Say something funny.

- What is your favourite quote?

- What are the goals of the Company?

  Safety, Cost, Efficiency, Customer Care (in whatever order is relevant)

- How are they measured?

  Number of accidents,% spend, Utilisation, Complaints (for example)

- What do you do about sexual harassment?

  Ask them not to complain (joke)

- You're in the restaurant on a stopover and in walks the Captain with a skirt on....what do you say?

  It depends on whether its colour clashes with his/her accessories (joke).

- Your career, though progressing nicely, is slow. How do you feel about it?

  Trick question - same as the one that asks if you would actually be happier in another company. You need to show confidence in yourself here.

- How would you rate yourself in relation to communication skills, dependability, and integrity on a scale of 1-10?

- If you attain all your goals as a pilot what do you intend to give back to aviation?

- What is important when managing pressure?

  Prioritisation, Ability to say No, Delegation, Asking for help.

- What are your salary expectations?

  You should have done some research into the averages. If you're still not sure, you could try turning it around and asking what's on offer. It will lower than they are prepared to offer because they want to keep their costs down so, if you are pressed for specifics, come up with a range of numbers rather than a single one.

- Why should we hire you rather than the previous candidate who has more hours?

- If you were a car (or anything else), what would you be?

  This is pressure to see how you will cope with the unexpected. It doesn't really matter what the answer is, but you will be expected to explain it.

- Why did you leave your last job?

  It's certainly not on to slate other companies or be too eager to leave your present one without a very good reason - if you can do either, you can do it to the one you're going for. Always remain positive about your current and previous employers because you never know when your paths may cross again. Besides, you will need a reference.

- What motivates you?

  Or, put another way: "What gets you out of bed in the morning?" Short of a huge salary, or a holiday home in Monaco, try for a constructive answer that will show what benefit you will bring to the business.

- What frustrates you?

  My answer to this one was "Not being allowed to do my best work."

- How would your former colleagues describe you?

  This is what's known in the sales trade as a buying signal - the thought of asking your previous employer for a reference is somewhere in their minds.

Don't forget you are new to the game, so you want standard answers with some common sense thrown in - if your responses don't seem to be what they want, back them up with good reasons why they should be - your training doesn't give you all the answers, but it does qualify you to think for yourself. Although an interview isn't meant to be funny, a couple of good, humorous remarks won't do any harm, but lay off the sarcasm - it doesn't go down well when you're supposed to be dealing with other people. The thing to remember is that all the above questions are based on fear (that you might screw up and make them look foolish, at the very least), so they are at a disadvantage, too. In large organisations, those who make mistakes don't get promoted - it follows, therefore, that people who don't make mistakes increase their chances markedly (you could also take the cynical view that those who do the least work make the least mistakes to its logical conclusion, but we'll ignore that for now).

Talking of which, if you were going for a management job, you might also want to consider:

- The work program is behind, the budget is overspent, and you are given some more work - how would you deal with it?

- How would you deal with someone who continually takes a sick day after returning from holiday or books dental appointments for late morning so they don't come back to work for another day?

- How would you deal with people who won't help colleagues who are returning from a hard day, saying "It's not my job", or "It's not my turn?"

Do not sit until invited, and if you are not, at least wait until the interviewer sits down. Do not smoke without permission, don't swear, interrupt or "interview" the interviewer, even if he is inept. Nor is it a good idea to argue, be familiar or apologise for yourself. The best tactic is to avoid extremes and place you and your opinions firmly in the middle - be the ideal "Company Person", in fact. By the way, the interviewers to watch out for are the surly or the quiet ones.

Don't even think of mentioning personality clashes or "philosophical differences" as they are more politely known (unless you want to be a trial lawyer!), and DO NOT TELL LIES.

Finally, when asked to do an evaluation ride, don't push off afterwards without helping to put the machine away, or at least offering to help.

There are many companies that are willing to invest in the right people (including the ones I used to run). From a seasoned Chief Pilot, "here are some of the lacking qualifications from 100 hour wonders we have seen in the last 3 seasons that led to their dismissal":

- Lack of ability to do basic math WITH a calculator - 2+2 =5 really?

- Unable to show up on time or on the right day.

- After hounding me for a couple months for a job and 2 days of exams, quit to go to work on a ski hill after 3 days of work.

- Ran out of gas in the company truck 200 kms short of their destination.

- "Dude I sold my car and bought a dirt bike but now I have to ride the greyhound so give me lots of notice for work man!!!"

- "I really don't care if I work because my gramma paid for my license"

- "See you at the hotel" as the engineer and captain continue to service the aircraft for another 4 hours.

- Won't remove their sunglasses even at night.

- Using the phrase "I'm a pilot, I don't have to do that." more than 3 times daily.

- "I'm not staying in that hotel" (even though the captain and engineer are quite happy).

In the last 3 years we have interviewed and/or hired and fired 14 100 hour pilots. Maybe the problem is not the amount of Canadian low timers but the quality. We now have a good CANADIAN low timer that we are in the process of beginning training and have trained 6 in the history of the company. I'm not surprised at the raising of minimum hours there are some numbskulls out there.

WHAT TYPE OF COMPANY?

Now you've got that job, it will all likely feel very strange. This chapter is here to help get you oriented! It describes the types of company you might be working for, who you might meet, with a look at some of the paperwork, and how to start your own company when the time comes.

As you probably already know, there are three types of flying, *Commercial Air Transport*, *Aerial Work* and *Private*. Being a professional (well, potentially, anyway), you will only be concerned with the first two.

- *Commercial Air Transport* exists where payment (usually by a passenger) is given for the use of an aircraft, which in this context means like a taxi, as opposed to self-drive car hire

- *Aerial Work* covers other situations where payment is still given, but in specialised roles not involving the usual passenger or freight carrying, such as photography or flying instruction, slinging, powerline survey, or any other situation where you're getting paid to fly

- *Private Flying* speaks for itself, its most distinguishing feature being that no payment exists, other than by the pilot, for the right to use the aircraft in the first place, although this in itself could cause problems

Within the above limits, the companies you could get involved with will also fall (broadly) into three categories, with some blurring in between, in the shape of *Scheduled, Charter* or *Corporate*.

Scheduled

"Scheduled Flying" is a legal definition describing services that run at predefined times with certain conditions imposed on them, such as being open to all classes of passenger and the flights always running, even though they may be empty. This would mean that, although oilfield helicopters do indeed move at predefined times, they are not subject to the other restrictions and are not therefore "Scheduled", but the difference is mostly transparent.

Scheduled Flying is said to be boring (actually it is), but it does have the advantage of being organised anything up to 4 weeks in advance, so you can at least have some sort of planning in other areas of your life; this is strictly enforced by the authorities, and is covered more in *Flight Time And Duty Hours*, below. Well, at least you know when you're going, even if you're not sure when you'll be coming back!

Charter

If scheduled flying is like bus driving, then charter flying is a taxi service, which means you are on call twenty-five hours a day with everything geared to an instant response to the customer, leaving you unable to plan very much. Don't get me wrong; this can be fun with plenty of variety and challenge in the flying, but the downside is an Ops Department that lets you do all the work yourself, and being left hanging around airports or muddy fields while your passengers are away (with missed meals, getting home late, etc.). Charter Flying is also where your other

skills as Salesman and/or Diplomat come into play, as you will be very much involved with your passengers, who are more than just self-loading freight!

Thus, while you can move relatively easily from Charter to Scheduled, it's not so straightforward the other way round. As a scheduled pilot, you rarely see your passengers, and the flying is very different. Charter (or Air Taxi) is intensive, single-handed and stressful work in the worst weather (you can't fly over it) in aircraft with the least accurate instruments.

As a pilot, you can have two types of working day, depending on the flying you do. In Scheduled, there is relatively little to do before departure as a lot is done by others - for example, ground staff check-in and weigh the passengers whilst engineers look at the aircraft, although you still need some knowledge of what they do, as the buck stops at the bottom. A day flying charter, however, is a different story. You could be working at almost any time, provided the Duty Hour limits are not exceeded (again, see below). Departures are inevitably very early, as businessmen need to be where they're going at approximately the start of the working day and return at the end of it, so some days can be very long.

As you're only allowed a certain number of hours on duty, there's a continual race to minimise them, sometimes working like a one-armed paper-hanger to keep up with everything. The flight plan has to be filed, the weather checked (as well as the performance and the aircraft itself), the passengers' coffee and snacks must be prepared and they must be properly briefed and looked after (that's just the start).

Usually, the only thing that can usefully be done the day before is to place the fuel on board, and even that can be difficult if the aircraft is away somewhere else. The flight itself is busy, too. As it's single-pilot, you do the flying, navigation and liaison with ATC. By contrast, the time at your destination is very quiet - after you've escorted your passengers through security and seen them safely on their way (the terminal's naturally miles away from the General Aviation park) you have to walk back to tidy up, supervise the refuelling, do the paperwork and have your own coffee (if there's any left) while preparing for the return journey.

If you're in a place you haven't been to before, you could always see the sights, but airfields are usually well away from anything interesting, with very few buses to get you there anyway. After a while, all you remember will be the same shops, so the general thing is to join the rest of the "airport ghosts", or other pilots in the same boat as you, and find a quiet corner to read a book. You may as well go to the terminal, because you have to meet your passengers there, but constant announcements could drive you out to the aircraft again.

However, while you may be on time to meet them, your passengers will very rarely be on time to meet you!

In Charter, it's also a luxury to have more than one day off in a row, and those you do get are needed by law, or turn up by surprise where you don't fly if business is bad, even though you've still gone into the office (the normal routine is - if you don't fly, you're not on duty, but common sense dictates that, if you're in the office doing something that is traceable, such as doing exams with a date on them, you'd better put down the hours). Some companies don't allow any leave at all during Summer, which is the height of the busy season, and only a week at a stretch if you do get it.

Corporate

Corporate flying, where you run the Flight Department for a private company, is similar to Charter, but not Commercial Air Transport, so the requirements (and paperwork!) are not so strict. Having said that, most corporate Flight Departments are run to Commercial standards, or better, and there is, naturally, no excuse for letting your own standards slip. One distinguishing feature is the way the Corporate world regulates itself - high performance intercontinental aircraft follow pretty much the same rules as single-engined General Aviation ones, and it's a credit to the people in it that things run so well.

In the Corporate world there are two types of Company. The first is the large conglomerate, where the aircraft is just as much a business tool as a typewriter is. You are genuinely a Company employee, people are used to the aircraft, you collect customers and move Company personnel around, from the Chairman to the workers, and your decisions as a professional are respected. There is a high degree of job satisfaction in this type of work, especially as you will build up relationships with regular passengers.

On the other hand, you might end up where the aircraft is the personal chariot of the Chairman, with you as its chauffeur (or, if you look at the books carefully, a gardener!), in which case nobody else gets to use it and what you think doesn't matter, because the sort of person who is dynamic enough to run a large company single-handed also thinks the weather will change just for him, and you're constantly under pressure to try and find the house in bad weather, which, naturally, hasn't got a navaid within miles. Unless you can establish a good personal relationship with your passenger, or have an extremely strong character, you are unlikely to get much job satisfaction here, especially if the company is family-run and you get to take the kids to horse shows, etc. at weekends. One CEO (now dead) used to insist that, when he was in the machine, no 2-minute rundowns were to take place.

Having said all that, there are some decisions that are not yours to take, whoever you work for. Unfortunately, you are only In Command where technical flying matters are concerned. If it's legal to fly then, strictly speaking, it's nothing to do with you whether it's sensible or not - it's an operational decision. If the Chairman (or Ops) wants you to fly and risk being left to walk if things get too bad, then it's entirely up to them - it's their money. For example, say you check the weather the night before and advise your passengers to go by car, because, while the destination and departure will be OK, the bit in the middle is iffy and there's no real way of knowing what it's like unless you go there and have a look.

However, they must get there and the timings mean they can't delay things till the weather gets better, so it's the car or flying - a straight choice. If your man wants to try and fly, and risks missing the meeting at the other end because you refuse to either start or carry on when it becomes impossible, then that, I suggest, is up to him. **Please note** that I'm not advocating flying in bad weather as a normal procedure! You might be able to fly in those conditions, but what happens later, when you can't find your way back (or meeting someone coming the other way!)

A major plus point about Corporate Aviation is the way companies spend money on their flagship. It's a curious fact that, despite the higher standards that Commercial Air Transport demands, I have never yet seen a badly maintained Corporate aircraft and very few badly run Corporate Flight Departments, but decidedly the opposite has often been the case in the commercial world. Corporate work sometimes pays the most, at least where smaller aircraft are concerned, but the jobs are less stable, as the aircraft is usually the first thing to go when the Company gets into financial difficulties (which is more often a wrong decision than you think). This often depends on how it is perceived by other parts of the organisation, so perhaps you could add marketing to your list of occupations.

COMPANY PERSONALITIES

The Company will appoint certain people to undertake particular tasks, and you will find some described below. Naturally, some will change, depending on your setup, and one person's functions may be combined with another's. Larger companies may swap the Ops Manager and Chief Pilot in terms of seniority. In Canada, the people described here must be qualified under *Commercial Air Service Standards* (CASS).

The Managing Director (or CEO)

This person has the ultimate responsibility for the efficiency, organisation, discipline and welfare of the Company, ensuring that all activities are safe and legal and that the Company is commercially viable. This will include marketing and projection of the Company image.

The Chief Pilot

Next in line is the Chief Pilot, who is the main point of reference that Inspectors and other officials will relate to, and they will expect to see him with some measure of control of the day-to-day happenings of the Company, although technically the job is just to keep things legal. However, to do that, there will have to be some involvement in the more commercial aspects (in Canada, this position, that is, next one down from the MD or CEO, may actually be occupied by the Ops Manager, for which see below).

The Chief Pilot is responsible to the Managing Director for the overall safety, legality, efficiency and economy of flying operations by the establishment of proper drills and procedures, and for ensuring that people (well, pilots, anyway) are properly qualified, so he will be responsible for hiring and firing.

Whilst the MD handles the administrative acceptability of work, the Chief Pilot has the technical side of things to worry about, like keeping control of the Flight Time and Duty Hours Scheme (sometimes by random inspection of returned flight documentation) in addition to supervising aircrew currency, maintaining aircrew records, compiling and updating the Ops Manual, raising occurrence reports and Flying Staff Instructions.

Randomly inspecting returned flight documentation is a real chore, and is done for three reasons; the first is that it's part of the Company's Quality Assurance Scheme, and the second is to ensure that you're doing your job properly. The third, and most important, is to eliminate nasty surprises when the Inspector drops in for coffee.

You will greatly endear yourself to your Chief Pilot if you make sure that *all* boxes on *all* forms are filled in (whether or not you think they're relevant), especially on the Technical Log, Loadsheet and Navigation Log (Plog), and not at the end of the day, because you might get ramp-checked before then.

Digression: When ramp-checking, Inspectors are looking for (amongst other things in a long checklist), altimeter settings, holes in the dashboard, approach plates out (or not), general condition of the aircraft, cleanliness, etc. and *scruffy paperwork*, with parts not filled in. They will especially be interested in Weight and Balance calculations.

With regard to the above items, where a signature is required, produce one, and always ensure that your departure fuel in the Tech Log agrees with the fuel load in the Load Sheet (all tanks) and the Nav Log, and that fuel usage throughout the flight is consistent with time, that is, that you're not using mysterious amounts of fuel that would indicate somebody's fiddling the books. Especially make sure that the fuel loads on the Tech Log and Loadsheet are above that required for the trip as specified on the Nav Log. The same rules apply to passenger and freight loads, and you should *always* check your figures, especially when adding up in hours, minutes and seconds - many engineers don't let pilots add up because it messes up the paperwork - they do all the entries themselves.

Lastly, don't write defects down on the Nav Log and forget to put them in the Tech Log at the end - that's a dead giveaway to your Inspector, as almost every aircraft goes unserviceable when it gets back to base as if programmed.

The Chief Pilot also liaises with the Maintenance Contractor on airworthiness matters, and may designate a suitable person within the Company to carry out, or be responsible for, any of the above duties. That person would be directly responsible to the Chief Pilot (as is everybody else). The Chief Pilot may also have the secondary function of:

Flight Safety Officer

Or FSO, who operates the Occurrence Reporting Scheme and maintains a vigorous Flight Safety policy, which entails collecting information from the various sources that publish it, and spreading it around the Company, probably by giving lectures and convening regular meetings with management, in accordance with the Quality System; this may also involve conducting internal investigations when somebody has an accident, and cooking up reports. The reason for spreading things

around is part of the reason for accident investigation, i.e. *that it doesn't happen again!* Safety management involves plenty of communication - as a Flight Safety Officer, you must encourage people to speak to you, so your personality is important. It's more than just a desk job.

The Chief Training Captain

This person coordinates flying training, arranges periodical checks and examinations, selects training staff, and ensures that flying training meets statutory requirements, if necessary by liaising with the Authorities, in addition to compiling and maintaining flying training records. Where Training Captains are thinly spread between companies, meaning that you don't see them from day to day, the Chief Training Captain may simply be the Chief Pilot wearing another hat, for consistency.

Base Manager

A sort of mini Chief Pilot/Ops Manager, in charge of a remote base, responsible to the Ops Manager or Chief Pilot for its day-to-day running and local marketing, keeping customers happy, altering your documentation, etc. In some companies, the Base Manager does the hiring and firing.

The Maintenance Contractor

The Maintenance Contractor maintains and valets Company aircraft in accordance with directions and laid down procedures or, more simply, mends what you bend. As to what laid down procedures is a good question, since they are also supposed to develop the Maintenance Schedule. If your Company does its own maintenance, you will find instead a Maintenance Manager and Chief Engineer, who will have to order spares and schedule maintenance in a timely fashion, together with everything else to run an efficient organisation.

Engineers

These are the guys who keep you up in the air, and it's not a good idea to upset them. They start work when you stop, often late in the night so you can fly next morning. Not only that, they have no duty hour regulations, so anything you can do to help is greatly appreciated, particularly cleaning aircraft. In fact, engineers also do a lot of operational stuff out in the field, and a good one is worth his weight in gold (more valuable than a pilot, actually).

This is from the late Dave Williamson, a senior Canadian pilot: (thanks, Cap!):

"I've always found that the maintenance I received over the years has more to do with the work habits and ability of my engineer than what dictates came out of any Head Office. After all is said and done, it is basically what you the pilot and he, the engineer will put up with. I've seen pilots who will put up with an awful lot, but the engineer was damn good, a hard worker and knew the a/c darn well........and in that scenario it was the ENGINEER that kept that a/c producing money and making sure that it got what it needed, when it needed it. Turn that situation around and as good as the pilot might be, you got a can of worms and lots of down time and unhappy customers. Just give me a good a/c, a good engineer and my down days you'll be able to count on one hand for the whole summer. I'll also buy that poor s.o.b. all the beer he wants because while I'm in my room watching TV, he's swatting at horseflies, etc. etc. trying to make me look like a hero for the next day. I've also taken a/c from hangars and they never saw home for a year and when they finally did return to the hangar (barring and major component changes), they were in as good a shape or maybe even better than when they left. To me at least, the buck stops with the engineer that accompanies me and NOT with the DOM or company policy because a good engineer has certain standards that he won't compromise for NOBODY because his reputation rides on what he produces and will put up with. Other engineers and pilots quickly notice and take note of the hard-working and able engineers and the word spreads and helps him down the road.

So am I a huge admirer of engineers?........ I KNOW a huge portion of the reason I'm still here on Mother Earth is because of the engineer working on my a/c and his standards of excellence and not anyone back at some company headquarters. The buck stops with him and that's why when my a/c is on the ground HE'S "the Man" and if he says it's grounded or HE got a problem with parts or HQ, then HE gets 1000% of my back-up and I'd walk before going against him. So if somebody at some company wants to push or lean on my engineer concerning doing something improper to my a/c, they're messing with the wrong engineer, on the wrong a/c and driven by the wrong pilot to play those games with. That is why I have a huge respect for the good engineers that I've had and maybe someday before I pass on, I'll see them get paid as much as pilots with the same experience (God forbid eh?), because they earn it in spades. Anyone in charge of making sure MY a/c stays in the air and in one piece deserves a healthy paycheque."

From another two:

Always feed and water your engineer. A dry and hungry one gets cranky through the day and will likley not smile and greet you with the love you deserve when you come back to base with twisted up bambi bucket cables.

If you screw up, have a six pack in your hand when you tell him about it. Big screw-ups require more beer, and the requirement is doubled for screw-ups that require a code of silence.

However, when the pilot of a medium rips open the udder or dump-valve of a bambi bucket, no amount of beer is ever enough. The pilot of said aircraft is in deep du-du even if he owns a brewery. The only way to redeem himself is to offer, no, BEG to help fix said dump-valve and then buy beer in mass quantities for remainder of the tour. Failure to help repair dump-valve casts said pilot into a life of never-ending hatred by the engineer...There is no possible way to not be hated by the engineer,......EVER!

HOW TO OPERATE A HELICOPTER MECHANIC
By William C. Dykes

Taken from the Internet.

A long, long time ago, back in the days of iron men and wooden rotor blades, a ritual began. It takes place when a helicopter pilot approaches a mechanic to report some difficulty with his aircraft. All mechanics seem to be aware of it, which leads to the conclusion that it's included somewhere in their training, and most are diligent in practicing it.

New pilots are largely ignorant of the ritual because it's neither included in their training, nor handed down to them by older drivers. Older drivers feel that the pain of learning everything the hard way was so exquisite, that they shouldn't deny anyone the pleasure.

There are pilots who refuse to recognize it as a serious professional amenity, no matter how many times they perform it, and are driven to distraction by it. Some take it personally. They get red in the face, fume and boil, and do foolish dances. Some try to take it as a joke, but it's always dead serious. Most pilots find they can't change it, and so accept it and try to practice it with some grace.

The ritual is accomplished before any work is actually done on the aircraft. It has four parts, and goes something like this:

- The pilot reports the problem. The mechanic says, There's nothing wrong with it."

- The pilot repeats the complaint. The mechanic replies, "It's the gauge."

- The pilot persists, plaintively. The mechanic Maintains, "They're all like that."

- The pilot, heatedly now, explains the problem carefully, enunciating carefully. The mechanic states, "I can't fix it."

After the ritual has been played through in it's entirety, serious discussion begins, and the problem is usually solved forthwith.

Like most rituals, this one has its roots in antiquity and a basis in experience and common sense. It started when mechanics first learned to operate pilots, and still serves a number of purposes. It's most important function is that it is a good basic diagnostic technique. Causing the pilot to explain the symptoms of the problem several times in increasing detail not only saves troubleshooting time, but gives the mechanic insight into the pilot's knowledge of how the machine works, and his state of mind.

Every mechanic knows that if the if the last flight was performed at night or in bad weather, some of the problems reported are imagined, some exaggerated, and some are real. Likewise, a personal problem, especially romantic or financial, but including simple fatigue, affects a pilot's perception of every little rattle and thump. There are also chronic whiners complainers to be weeded out and dealt with. While performing the ritual, an unscrupulous mechanic can find out if the pilot can be easily intimidated. If the driver has an obvious personality disorder like prejudices, pet peeves, tender spots, or other manias, they will stick out like handles, with which he can be steered around.

There is a proper way to operate a mechanic as well. Don't confuse "operating" a mechanic with "putting one in his place." The worst and most often repeated mistake is to try to establish an "I'm the pilot and you're just the mechanic" hierarchy. Although a lot of mechanics can and do fly recreationally, they give a damn about doing it for a living. Their satisfaction comes from working on complex and expensive machinery. As a pilot, you are neither feared nor envied, but merely tolerated, for until they actually train monkeys to fly those things, he needs a pilot to put the parts in motion so he can tell if everything is working properly. The driver who tries to put a mech in his "place" is headed for a fall. Sooner or later, he'll try to crank with the blade tied down. After he has snatched the tailboom around to the cabin door and completely burnt out the engine, he'll see the mech there sporting a funny little smirk. Helicopter mechanics are indifferent to attempts at discipline or regimentation other than the discipline of their craft. It's accepted that a good mechanic's personality should contain unpredictable mixtures of irascibility and nonchalance, and should exhibit at least some bizarre behavior.

The basic operation of a mechanic involves four steps:

1. Clean an aircraft. Get out a hose or bucket, a broom, and some rags, and at some strange time of day, like early morning, or when you would normally take your afternoon nap) start cleaning that bird from top to bottom, inside and out. This is guaranteed to knock even the sourest old wrench off balance. He'll be suspicious, but he'll be attracted to this strange behavior like a passing motorist to a roadside accident. He may even join in to make sure you don't break anything. Before you know it , you'll be talking to each other about the aircraft while you're getting a more intimate knowledge of it. Maybe while you're mucking out the pilot's station, you'll see how rude it is to leave coffee cups, candy wrappers, cigarette butts, and other trash behind to be cleaned up.

2. Do a thorough pre-flight. Most mechanics are willing to admit to themselves that they might make a mistake, and since a lot of his work must be done at night or in a hurry, a good one likes to have his work checked. Of course he'd rather have another mech do the checking, but a driver is better than nothing. Although they cultivate a deadpan, don't-give-a-damn attitude, mechanics have nightmares about forgetting to torque a nut or leaving tools in inlets and drive shaft tunnels. A mech will let little gigs slide on a machine that is never pre-flighted, not because they won't be noticed, but because he figures the driver will overlook something big someday, and the whole thing will end up in a smoking pile of rubble anyway.

3. Don't abuse the machinery. Mechanics see drivers come and go, so you won't impress one in a thousand with what you can make the aircraft do. They all know she'll lift more than max gross, and will do a hammerhead with half roll. While the driver is confident that the blades and engine and massive frame members will take it, the mech knows that it's the seals and bearings and rivets deep in the guts of the machine that fail from abuse. In a driver mechanics aren't looking for fancy expensive clothes, flashy girlfriends, tricky maneuvers, and lots of juicy stories about Viet Nam. They're looking for one who'll fly the thing so that all the components make their full service life. They also know that high maintenance costs are a good excuse to keep salaries low.

4. Do a post-flight inspection. Nothing feels more deliciously dashing than to end the day by stepping down from the bird and walking off into the sunset while the blade slowly turns down. It's the stuff that beer commercials are made of. The trouble is, it leaves the pilot ignorant of how the aircraft has fared after a hard days work, and leaves the wrench doing a slow burn. The mechanic is an engineer, not a groom, and needs some fresh, first hand information on the aircraft's performance if he is to have it ready to go the next day. A little end-of-the-day conference also gives you one more chance to get him in the short ribs. Tell him the thing flew good. It's been known to make them faint dead away.

As you can see, operating a helicopter mechanic is simple, but it is not easy. What it boils down to is that if a pilot performs his pilot rituals religiously in no time at all he will find the mechanic operating smoothly. (I have not attempted to explain how to make friends with a mechanic, for that is not known.)   Helicopter pilots and mechanics have a strange relationship. It's a symbiotic partnership because one's job depends on the other, but it's an adversary situation too, since one's job is to provide the helicopter with loving care, and the other's is to provide wear and tear. Pilots will probably always regard mechanics as lazy, lecherous, intemperate swine who couldn't make it through flight school, and mechanics will always be convinced that pilots are petulant children with pathological ego problems, a big watch, and a little whatchamacallit. Both points of view are viciously slanderous, of course, and only partly true.

HUMOUR

Engineers can also have a sense of humour - here are some of their entries in response to pilot complaints and problems, generally known as squawks:

* Left inside main tyre almost needs replacement. *Almost replaced left inside main tyre.*

* Test flight OK, except autoland very rough. *Autoland not installed.*

* No. 2 propeller seeping prop fluid. *No. 2 propeller seepage normal - Nos. 1, 3 and 4 propellers lack normal seepage.*

* Something loose in cockpit. *Something tightened in cockpit.*

* Evidence of leak on right main landing gear. *Evidence removed.*

* DME volume unbelievably loud. *Volume set to more believable level.*

* Friction locks cause throttle levers to stick. *That's what they are there for!*

* IFF inoperative. *IFF always inoperative in OFF mode.*

* Suspected crack in windscreen. *Suspect you're right.*

- Number 3 engine missing. *Engine found on right wing after brief search.*

- Aircraft handles funny. *Aircraft warned to "Straighten up and Be Serious."*

- Target radar hums. *Reprogrammed target radar with words.*

- Mouse in cockpit. Cat installed.

The Operations Manager

Although this person may be technically under the Chief Pilot, in practice, they have more or less equal status and, in some companies may have one person occupying both positions. Having said that, Ops have to acknowledge your ultimate authority as aircraft commander. In addition, where Ops Managers must have certain qualifications, such as in Canada, and may therefore have more than the Chief Pilot anyway, you may find that the Ops Manager is well and truly in charge and the Chief Pilot a few steps down in the pecking order. Look for this situation in larger companies, where you will also find Ops Assistants doing most of the work described below.

Operations will provisionally accept work and, in liaison with the Chief Pilot, confirm it. This means that they organise the flying program, including pilot duty and rest days, so you want to keep on their good side. Ops will ensure that duty times are in limits by keeping a record of flight crew flying and duty hours, and are supposed to ensure that you receive a written briefing (including NOTAMs, etc.) before going anywhere, and that all passenger and cargo manifests and tickets are completed as required.

The Ops Manager must keep in touch with the Maintenance Contractor to ensure scheduling for maintenance, forwarding completed Tech Log sheets and other relevant documents to them at the end of each flight. This is not the same as mentioned for the Chief Pilot, who does it on a more lofty level - all the Ops Manager is expected to do is monitor the aircraft hours so that nothing gets behind, and everything gets serviced on time. This is usually done by circulating coloured copies of the Tech Log after a flight.

Operations will also maintain carnets and aircraft documents (collectively referred to as aircraft libraries), an up to date stock of maps, route guides and aeronautical charts covering all areas of Company operations, Flight Information Publications (such as NOTAMs, Air Pilot, AICs, Royal Flights, the Landing Site Register, etc.), and arrange exemptions and clearances for particular tasks.

Note: Although Ops are supposed to ensure the validity of all licences, medicals, periodical checks and training, you still have to keep your own up to date.

The Ops Manager also ensures that Company accident and incident procedures are followed, processes amendments of the Operations Manual, assesses landing sites, categorises airfields, calculates specific weather minima, obtains met forecasts for planned routes and destinations, and arranges overnight accommodation for night stops, amongst other things. Most important is the arrangement of an accurate flight watch of all company aircraft movements and a standby telephone coverage outside normal working hours. This is not legally required under some circumstances, such as Day VFR in Canada, but is still Good Practice.

A company that actually gets the Ops Manager to do all that is setting quite a high standard (naturally, the above duties may be delegated). Unfortunately, what happens is that whoever owns the Company has a nephew, niece, girlfriend or whatever, who ends up doing the job instead. In that case, the best thing you can do is either leave the company, or this book around! In Canada, Ops Managers must hold, or have held, a pilot's licence for one of the types flown, or have appropriate experience, and rightly so. While on the subject, the biggest thing you need out of Ops is information, so try and make sure they get it from the customer, or you will continually find yourself having to fix other peoples' problems illegally, as when you turn up for a sling job expecting a 200 lb load and find it's actually nearer 600, which means pressure on you to go overweight to get the job done.

Quality Assurance/Compliance Manager

The Compliance Manager (who may be the Chief Pilot in disguise) ensures that the company's quality system is established and implemented. Duties include the issue and withdrawal of documents and forms, and maintaining a list of them, together with the aforementioned regular checks of documentation, etc. Routine flights should also be accompanied occasionally to confirm that normal procedures are being followed, but this will likely take the shape of a Training Captain doing a Line Check.

The Company Pilot

In small companies, it will be policy to operate on a single crew basis as far as possible (less wages to pay), with the designated Commander occupying the Captain's seat as per the Flight Manual. It's therefore important to maintain your own standards, because you'll be on your own a lot.

You may think it a little over the top to see somebody with large amounts of gold braid emerging from a small aircraft and be wondering on what occasions you can call yourself "Captain". As far as I can make out, it used to be a convention that if you had either 5000 hours, an ATPL of some description or a Training (IRE) qualification, you were entitled to do so. The trouble was that as smaller airlines became popular, they didn't have people so qualified and passengers were wary of flying with pilots who didn't have the shiny stuff on their sleeves. Thus, the various rank gradings have become blurred and you're a commander (small c) if you're in charge of any aircraft, in the same way that people in charge of smaller seagoing vessels are Ship's Masters, as opposed to Captains.

You may also be wondering why the commander has to be designated - this is so the Subsequent Board of Inquiry can pin the blame on the right person. In the USA, for example, under certain circumstances, four people can claim PIC time, including those in the passenger seats! This would naturally include whoever is doing the poling, but if someone in another seat has better qualifications, or is the owner or operator, that would qualify, too, especially in court (so watch it). Certainly, in the military, the Captain has never been necessarily the first pilot (it could easily be a senior officer in a passenger seat), since the Captain is responsible for the final disposition of the aircraft (which, when you think about it, could also include a purser). So, there is a difference between acting as PIC and logging PIC time, and it should be spelt out clearly to save legal (and CRM) trouble later - in fact, having two Captains on board, with neither sure of who's in charge can be a real problem. Either they will be scoring points off each other, or be too gentlemanly, allowing an accident to happen while each says "after you". If it comes to that, how do you sort out the mess if you have someone in the left seat who is a First Officer pretending to be a Captain, and someone in the other seat who is a Captain pretending to be a First Officer?

Anyhow, as an aircraft commander, you are first and foremost subject to any aviation regulations that may be in force. Inside the Company, you are responsible to just about everybody else (but especially the Fleet/Base Manager or Chief Pilot) for ensuring that aircraft are flown with prime consideration for the safety of passengers and persons on the ground; not negligently or wilfully causing an aircraft to endanger persons or property while ensuring it is operated in accordance with performance requirements, Flight Manuals, checklists, State authority regulations, the Operations Manual, Air Traffic Regulationsand NOTAMs.

Seems a bit much, doesn't it? Hang on.......

It's also up to you to keep your licences and personal flying logbooks up to date, and to ensure you are medically fit for your duties (a Board of Inquiry or insurance company may interpret the words "medically fit" a little differently than you think if you fly with a cold or under the influence of alcohol). You must keep customers and the Company informed of any accidents, incidents and alterations caused by bad weather or other reasons. In remote areas, this will include a position report every hour or so. Yours is the final responsibility for supervising the loading, checking and refuelling of your aircraft and making sure that all passengers are briefed on Emergency Exits and the use of safety equipment (see later), although you also have the right to exclude certain persons, such as drunks, etc.

You must check that the aircraft is serviceable with a current Certificate of Release to Service (or equivalent) and with previously reported defects noted in the Technical or Journey Log as being rectified or transferred to the Deferred Defects lists by a person so qualified. Any defects must be allowed for in the Minimum Equipment List (MEL) or CDL.

You must ensure that no weight limitation is exceeded, that the C of G will remain inside the envelope at all times, and that performance is sufficient to complete the flight, as well as leaving a duplicate copy of the Loadsheet and Technical Log (or Operational Flight Plan, in Canada) with a responsible person before each flight, and ensuring that all documents are correctly completed and returned to Ops at the end (all documentation must remain valid throughout the flight). Of course, nobody ever does this, but you are supposed to.

You should not permit any crew member to perform activities during takeoff, initial climb, final approach and landing that are not required for safe operation, and take all reasonable steps to ensure that, before take-off and landing, the flight and cabin crew are properly secured in their allocated seats (cabin crew should be secured in their seats during taxi, except for essential safety related duties).

Whenever the aircraft is taxying, taking off or landing, or whenever you consider it advisable (like in turbulence), you should ensure that all passengers are properly secured in their seats, and cabin baggage is stowed.

In an emergency situation (that is, requiring immediate decision and action), you should take any action considered necessary under the circumstances, which means you can break all the rules in the interest of safety.

You can apply greater margins to minima at any time.

You should ensure that a continuous listening watch is maintained on relevant radio frequencies at appropriate times, which, officially, is whenever the flight crew is manning the aircraft for the purpose of commencing and/or conducting a flight, and when taxying.

You should not permit a Flight Data Recorder or Cockpit Voice Recorder to be disabled, switched off or erased, especially after an incident or accident, unless you need to preserve what's on the CVR (because it erases automatically as power is reapplied).

Although it's part of Operations' job to get a met forecast, it's actually your responsibility, so you may as well do it yourself.

Your behaviour and representation of the Company in front of actual and potential clients must be exemplary.

Finally, here's a little gem, from about 1919, which comes from *Recollections of an Airman*, by Lt Col L A Strange. Nothing changes!

> *"…As a pilot of a machine, you are responsible for that machine all the time, and it is always your fault if you crash it in a forced landing occasioned by any failure, structural or otherwise, of the machine or its engine. It is your fault if, in thick weather, you hit the top of any hill that has its correct height shown on your map.*
>
> *It is entirely your fault if you run out (of petrol) when coming home against a headwind after four or five hours (of flying), or if you fail to come down on the right spot after a couple (of) hours cloud flying.*
>
> *It is your fault if you have nowhere to make a landing when the engine fails just after you have taken off; in the event of a forced landing, your machine is a glider that should take you down safely on any possible landing place.*
>
> *It is your fault - well, it is a golden rule to assume that whatever goes wrong is your fault. You may save yourself a lot of trouble if you act accordingly."*

The First Officer

First Officers must know of the duties and responsibilities of the commander in case of incapacitation, so they will more than likely find themselves preparing and maintaining the navigation and fuel logs in flight, because they should be fully aware of the intended route, weather, etc. that may affect it. Constant briefings from the commander are essential, as the FO naturally must know the game plan if there is going to be a takeover at any stage. This even extends to the routes to be flown,

minimum safety altitudes, overshoot action, etc. All this "interaction" is part of Crew Resource Management, of which more elsewhere. In addition, First Officers carry out checks (the commander reads them, or vice versa), make radio calls, cross-check altimeters and other instruments and monitor each flight.

They're supposed to advise you (as Commander) of any apparently serious deviations from the correct flight path, such as specific warning if, on an instrument approach, the rate of descent exceeds 1000 feet per minute or the ILS indicator exceeds half-scale deflection, or of any instrument indicating abnormal functioning, which is difficult with a lot of head-down work of their own to do. In addition, they carry out secondary checks on engine power after the throttles have been set. If, for any reason, you become incapacitated, they should be prepared to assume command. They also supervise loading and refuelling and prepare the loadsheets for the Commander's signature before each flight, if it's not already done by a handling agent. When it's raining, they do the preflight check.

Others

There may well be other staff around, such as Flight Despatcher, Flight Follower, Ramp Officer, Senior Steward(ess), etc. who are not catered for here, but it shouldn't be hard to deduce what they get up to, given the above examples. There isn't a specific qualification for Flight Despatchers in UK, either, but, sometimes, they get a whole week's training.

However, let's not forget:

Customers

These are the most important people in any company, for obvious reasons, but different parts of the industry allow them a greater or lesser degree of freedom in dictating how the job is done (we're talking about air taxi or aerial work here), which will range from specifying the number of hours pilots will have (fair enough) to insisting that they shouldn't wear a seat belt or that you pull torque over the limits to get the job done (unacceptable). Please do not get me wrong - most customers are entirely reasonable and rely totally on your judgement as a professional, but there are some with enough knowledge to be dangerous, who have no respect for your position and are the ones who need the most tactful handling. It's easy to say that you don't need that type of customer anyway, but money is money, so what works (for me, anyway) in those situations is just to say you don't feel comfortable doing whatever they ask, and suggest an alternative (most important).

Don't explain why, it just confuses the issue and gives them something else to hang you with when they complain to head office later. They won't get the story right anyway, so don't make it worse.

I have found that most of them are testing you to see how far they can go, as they would in any other walk of life, so lay down the law straight away. Of course, you may get run off the job, but most will respect your limits after that.

THE OPERATIONS MANUAL

Apart from the people above, almost the first thing you might see in your new company will be the Operations Manual. This is usually fairly badly written, often being a copy of somebody else's, which will no doubt include their bad English ("acquiring" Ops Manuals is a favourite form of Industrial Espionage). You'll probably also find items in the most illogical places, after being added willy-nilly over the years with no thought to content. It might also have been typed by someone wearing boxing gloves. It wouldn't be so bad if you were given time to read it, but you're usually expected to do so overnight, at the same time as learning the rest of the Company procedures and studying for the exams you will no doubt be expected to sit the following morning (as you've discovered already, everything happens yesterday).

The Operations (or Ops) Manual is like the Standing Orders issued by any military unit, hospital or other type of large organisation. It's a book of instructions that are constant, so that Company policy can be determined by reference to it, containing information and instructions that enable all Operating Staff (i.e. you) to perform their duties. It's partly to save you constantly pestering Those On High, but mainly for situations where you can't speak to them anyway and need information with which to make decisions. As part of the Operating Staff of a Company, you are subject to the rules and requirements in it, and it's your responsibility to be fully conversant with the contents at all times. You will be expected to read it at regular intervals, if only because it gets amended from time to time.

The Chief Pilot is usually responsible for the contents and amendment policy (he may well have written it as well, so be careful when you criticise the English). Amendments, when they're issued, consist of dated and printed replacement pages on which the text affected is marked, ideally by a vertical line in the margin. On receipt of an amendment list, those responsible for copies of the

manual are supposed to incorporate the amendments in theirs and record it on the form in the front, although when they leave the company, you usually get their copies back with nothing actually inserted, although the amendments might be signed for. You should find a proposal form for changes somewhere as well.

The prime objective for the Ops Manual being written in the first place is to promote *safety* in Company flying operations. As the authorities are involved, it's therefore compiled in accordance with the law (in fact, as far as you are concerned it is the law) and all flights should be conducted to the Commercial Air Transport, standards set out in it. There should be a definition of the phrase which, officially, is an aircraft operation involving the transport of passengers or cargo for remuneration or hire, which does not include Aerial Work or Corporate Aviation. Also, there will be a declaration of who you're working for, which may sound daft, but many companies trade under several aliases, and they will be pinned down as to their real identity somewhere in the first few pages.

Some parts of an Ops Manual apply even when you think you're flying privately, because the aircraft will still be operated by an *air transport undertaking*. There should be an indication of what bits relate to what types of flight, but most companies apply the same rules to everything, as it makes life easier. Usually consisting of several parts, the Manual can be the size of a single volume with a small operator, or several in the average airline.

The separate parts will consist of:

Part	Contents
A	The main volume, with admin and operating policy
B	Flight Manuals for each type operated
C	Flight Guides (Jeppesen, Aerad, etc., or even your own)
D	Training Manual

Manuals are notoriously difficult to navigate around, so a good index is important, as is a table of contents. This helps two people; you, trying to find the answer to a question in a hurry and the Inspector when reading the thing in the first place.

Although the manual will be supplemented by statutory instructions and orders, not all of them will be mentioned. It doesn't mean that you should ignore those that aren't, but being acquainted with all regulations, orders and instructions issued by whoever is all part of your job.

Naturally, references made to any publication (such as Air Navigation Orders of whatever year) should be taken as meaning the current editions, as amended. When they are mentioned in the Ops Manual, they acquire the same legal force.

There will be several copies of the Operations Manual around, the numbers issued differing with the size of the Company, but the typical distribution list below should be regarded as a minimum; each aircraft will have its own copy. All must be clearly marked for amendment purposes, and there's no reason why you can't have small versions for small aircraft, but remember they must all have the same text.

Copy	Who has it
1	Master Copy-Operations Manager
2	Relevant Authority (FAA, CAA, Transport Canada)
3	Chief Pilot
4	Training Captain
5	Maintenance Organisation
6+	One per aircraft or pilot

A large outfit will likely have its own print shop just for Ops Manuals and amendments.

FLIGHT TIME & DUTY HOURS

You have a maximum working day laid down by law, intended to ensure you are rested enough to fly properly. It's similar to truck drivers' hours, except that there's no tachograph; companies and pilots are trusted to stick to the Ops Manual and the authorities reserve the right to spot check the paperwork at regular intervals, mainly looking to see that flights are planned within the Company's scheme (if you don't see an Inspector for long periods, then you can assume that your Company is well regarded in this respect).

In UK, at least, *these regulations do not apply elsewhere*, so corporate pilots (or unpaid instructors) have no protection, apart from any basics under the regulations. Consequently, you could find yourself in continuing battles with Company executives, to whom working 28 12-hour days non-stop is not uncommon - if you are moonlighting from the military, you must count that time as well. It's fair to point out, though, that it's difficult to introduce Duty Hours into a corporate environment - the schedule changes so often that you would need a lot of extra staff to cope with it. I suppose you could point out that if the aircraft is not flown under the regulations, the insurance becomes invalid.

In some countries, such as Canada and USA, knowledge of flight time limitations is actually part of the Commercial Pilot's exams, whereas in UK you don't really start finding them out until you join a company.

Put simply, there is a basic working day, which generally is 10 hours long (14 in Canada). In UK, this may be longer or shorter, depending on the time you start and the number of crew you have; the earlier you start, the less time you're allowed. Within the resulting Duty Period there may be a maximum number of flying hours which cannot be varied, such as 7 hours' helicopter flying within 12 on duty. If you need an exceptionally long working day, you can always apply for an exemption to cover it.

The reason why the duty day is so lng in Canada is because the flying season is short, especially in the North, and full use must be made of the time available. That doesn't mean you can abuse the rules, though - there are stiff fines for breaking the limits, for companies and pilots, so beware. You can do as many hours as you want in the basic 14-hour day, limited only by refuelling, eating, etc., but if you extend to 15 hours, you can only do 8 hours flying.

Your Responsibilities

These stem from various provisions of the regulations, such as ANO and CAP 371 and/or Part Ops (Europe), or CARs in Canada. Firstly, you must inform the Company of all your flying (including Aerial Work, which in turn includes paid flying instruction), except (in UK) private and military flying in aircraft under 1600 kg MAUW. Exempt military time must be for the RAF Cadet Organisation, or experience flights. Anything else counts.

It's also up to you to make the best use of any opportunities and facilities for rest provided, and to plan and use your rest periods properly - you should inform Operations if you can't sleep properly, for example, who might arrange a specialist. Then there's the Aircraft Crew and Licensing part of the ANO (CARs in Canada) which says that you're not entitled to act as a member of a flight crew if you know or suspect that your physical or mental condition renders you temporarily unfit to do so (what about if you're permanently unfit?). In short, all this means you should not act as a crewmember (and should not be expected to) if you believe you are (or likely to be) suffering from fatigue which may endanger the safety of the aircraft or its occupants.

Company Responsibilities

Duties must be scheduled within the Company's approved scheme, and rostering staff must be given adequate guidance. Work patterns must be realistic with the intention of avoiding, as far as possible, over-running limits. As a result, they must avoid such nasties as alternating day and night duties and positioning that disrupts your rest. Not that they ever do.

Unless you're in an airline or offshore, it's obviously difficult to schedule much in advance, but companies must advise you of work details as far ahead as they can (though not less than 7 days), so you can make arrangements for adequate and, within reason, uninterrupted pre-flight rest. Away from base, it's normally the Company's job to provide rest facilities (the legal definition is "satisfactory in respect of noise, temperature, light and ventilation"), but they may lumber you with finding them, as you're on the spot - they are allowed to claim that short notice precludes them doing it. Note that bush and fire-fighting camps must also meet the definition of "suitable accommodation", which, in Canada, at least, should start as a single-occupancy bedroom subject to minimum noise, but can change to something suitable for the site and season, so you can't win. On seismic support, the only place you could get to hide is inside the back of your helicopter, which the European authorities don't allow.

All this being said, it must be pointed out before we go on that very, very few companies below a certain level are actually honest about their duty hours. The reason is fairly simple in most cases - if you kept to the letter of the law you would duty-hour yourself out of business, especially when there's not a lot of staff around (it's no different for truck drivers). Otherwise, the companies simply have no respect for the law or their employees; many are cheapskates and beat down on the room price, so you will get the noisiest and hottest for "suitable accommodation". Discretion here is the better part of valour, but falsifying duty hours is but a short step removed from doing it to other documents, and that would never do. It's hard to give you any advice, except to point out that being pedantic can often be counter-productive - that is, sometimes you just have to swallow things.

Maximum Duty Period (FDP)

A *Duty Period* is any continuous period through which you work for the Company, including any FDP (see below), positioning, ground training, ground duties and standby duty. A *Flying Duty Period* (FDP), on the other hand, is any duty period during which you fly in an aircraft as crew. It includes positioning immediately before or after a flight (say in a taxi or light aircraft) and pre/after-flight duties, so the start will generally be at least 30 minutes before the first scheduled departure time and the end at least 15 minutes after rotors last stopped time, though these may vary between companies.

Discretion to Extend FDPs

There are always delays in aviation, for anything from technical to weather reasons, and a Flying Duty Period may be extended if you think you can make the flight safely and have consulted the other members of the crew about their fitness. However, the normal maximum is based on the original reporting time, and calculated on what actually happens, not what was planned to happen (everything must be planned properly). Sometimes, for example, you may have to exercise it if a lower performance aircraft is used instead of a larger one, and consequently takes longer to get round the route. This discretion is yours (as Commander), but some Companies will make the decision for you before the first flight of the day, which is not when it should be used. Whenever discretion is exercised, the circumstances should be reported to Ops on the Discretion Report Form (in the Ops Manual).

If this page is a photocopy, it is not authorised!

In Canada, most people extend to 15 hours, if the following rest period is extended by the same amount (1 hour) or you do less than 8 hours' flying a day. Transport Canada can issue a special permission for 15 hours, so check your Ops Manual - in some cases, your next rest period must also be an hour longer, or you can't do more than 8 hours flying the next day. Spray pilots are restricted to 14 hours anyway (and must have 5 days off in every 30).

Minimum Rest Periods

As well as having a maximum number of hours on duty, there's also a minimum rest time between duty periods. A Rest Period is time before a Flying Duty Period which is intended to ensure that you're adequately rested before a flight. It doesn't include excessive Travelling Time (over 90 minutes or so) or Positioning. During it, you should be free from all duties, not interrupted by the Company and have the opportunity for a minimum number of consecutive hours sleep (8, in Canada) in suitable accommodation, plus time to travel there and back, and for personal hygiene and meals.

In Canada, the minimum rest period is defined in only one place in CARs, right at the front, under Interpretation, where it says that you should be free from all duties, not be interrupted and be able to get at least eight hours' sleep in suitable accommodation, travel there (and back) and take care of personal hygiene. Realistically, therefore, the rest period should be about nine hours long. It's an hour extra anyway for spraying, with 5 hours of sleep taking place between 2000 and 0600 hours. Time spent on essential duties required by the Company after duty are not part of any rest period.

You should have your rest periods (see also Duty Cycles) rostered enough in advance to get your proper rest. Minimum Rest Periods should be at least as long as the preceding Duty Period, and at least 12 hours (in UK), except when accommodation is provided by the Company, in which case the minimum may be 11, subject to any exemptions you have. Rest starts from the end of the Duty Period and not the Flying Duty Period.

Discretion to Reduce a Rest Period

You can reduce Rest Periods below the minimum, but like extending Duty Time, it's at your discretion, and can only be done after consulting the crew. In any event, you must be able to get at least 10 hours at the accommodation where you take your rest, subject to the requirements of Travelling Time. Use of discretion for reducing rest is considered exceptional and shouldn't be done to successive Rest Periods (it's very much frowned upon). In general, you're better off extending an FDP than reducing a Rest Period if at all possible. Also, at no time should a Rest Period be reduced if it immediately follows an extended Duty Period, or vice versa (this is even more frowned upon).

Split Duties

You can extend a duty day by other means than discretion, though, and you can do it on duties with a long time gap between flights. Technically, a Split Duty is a Flying Duty Period with two or more sectors separated by less than a minimum Rest Period, typically being a situation where you deliver people to a place and wait for them to come back. In other words, you can claim half of the period spent hanging around in the middle as "rest" and tack it on to the end of the basic working day. What's more, you can plan to do this from the start, extending the FDP by half of the "rest" taken if it's between 3-10 hours (inclusive, providing the hours are consecutive). In Canada, the extension is up to 3 hours, so you need 6 hours off.

In Canada, you can go beyond 14 hours by half the rest period up to 3 hours, *if* you have been given advance notice *and* you get 4 hours in suitable accommodation, *and* are uninterrupted. This means the maximum time you can possibly be on duty is 17 hours, if you have 6 hours off during the day. Your next rest period must be increased by at least the extended time.

Positioning

Positioning means being transferred from place to place as a passenger in surface or air transport, usually before or after a FDP, but also at any time as required by the Company (this shouldn't be confused with normal travel from home to work-see Travelling Time below). Many airlines use taxis for this, but you may be lucky and get a comfortable bus or a light aeroplane. All time spent on positioning is classed as Duty, and when it comes immediately before a Flying Duty Period is included in it, so the subsequent rest period must account for (and be at least as long as) the total FDP and positioning. Positioning is not, however, counted as a sector, and in case I haven't mentioned it, a sector is the time between an aircraft first moving under its own power until it next comes to rest after landing (there are no sector limits for helicopters).

SETTING UP A COMPANY

Most pilots are quite happy graduating to larger and larger types as their experience grows and don't concern themselves about operating their own aircraft. One day, though, there will be an opportunity to set up your own Company and obtain your own AOC, typically where you come across somebody with an aircraft who would like to offset the costs of operating it against some income. Or it may be that you come into some money yourself and feel able to go it alone. More common is where you fly someone on a charter who is new to flying, they become impressed and decide to buy their own aircraft, and because you were their first pilot you get made an offer.

There's nothing wrong with this, but think seriously before leaving employment with relative security for something that may only last a few months. One rule of thumb (which works very well) is that the more attractive the package offered, the less stable is the job. Another is to subtract twenty-five from the physical age of anyone who's keen on Aviation for its own sake, and wants to make a business out of it, to get their mental age. If you can, find out something about the company your prospective employer runs. Have they got credibility? Are they well established, or is the man you're talking to just an idiot with access to other people's money rather than his own? (See how many of his cheques bounce, and how much of what he says will happen actually comes true). Signs that a company won't last long include excessive flamboyance on the part of the Boss, who naturally pays for everything, treating all and sundry to lunches, drinks, etc. (if it was really his money, he wouldn't be doing that). Statements to the Press that are less than complimentary to other companies around should also be noted, as should excessive hype and illogical spending on non-essentials, where the basics aren't being looked after - that is, they spend money on smart new offices rather than servicing the aircraft.

I don't mean to put you off from anything that could lead to greater things, but a little scepticism in the early stages could save you and a lot of other people plenty of aggravation later on. There are several ways to protect yourself. One is, don't move at all - bide your time and see what actually materialises out of what your prospective employer promises. Many people say a lot, but not much actually happens - they give the illusion of movement without actually progressing anywhere (like 'reorganisation' in a large company). The more urgency projected, the more sceptical you should be.

Another way of protecting yourself is to have somewhere to go to if things fail to materialise; an even better way is to estimate the amount of work you can expect to do for a year and insist on payment of the whole lot in advance. If they want you enough, they'll produce the goods. In fact, to work in Aviation at any level (as opposed to playing in it) needs a more businesslike approach than most people think, at least as far as mental attitude is concerned. The whole idea is to earn money; if you happen to enjoy it, you're lucky, but you are doing yourself and other pilots a disservice by underselling yourself just because you're keen to fly. You suffer the same fate as companies who undercut - in the end, the waters just get stirred up, nobody makes any money at all and very few survive.

Financial Matters

If, despite the gloom, you still want to carry on, please let me add one more word of warning - you will need much, much more money than you anticipate. Not for nothing is it said that to end up with a small fortune in Aviation, you need to start with a large one! But it needn't be that bad, provided things are done properly from the start.

First of all, if you need to borrow money, you'll need as much slack as you can get to cover cash flow while you're waiting for customers to pay, and emergencies - if you only ask for just enough money, it will be patently obvious you don't know what you're doing and will be shown the door. When running an airline, you will find that major travel agencies can take up to 6 months (or more) to pay their bills, if they pay at all, which will cause major cashflow problems. Once the fuel companies don't give you any more credit, you don't have long to go, because people who owe you money definitely won't pay if they can get away with it.

Also, things only work out cheaper if you can afford to fork out the money from the start. Buying your own bowser, for instance, instead of positioning your helicopter to the local airfield for fuel, will probably cover all that empty flying and unnecessary landing fees inside three months, but you have to have the money in the first place - paying as you go along should be avoided as much as possible, as it will usually kill any project stone dead.

Don't depend purely on loans. In fact, you probably won't get one till the lender sees some input from another source (preferably yours), so you may need to find a Venture Capitalist who would be willing to invest in your project. These sort of people supply money in return for stock (shares in the Company), typically expecting to be free of their obligation in about four years or so with a handsome profit (although they could make a loss). The major benefit to you is that they provide ready cash and stability without your spending power being drained by interest payments. Although a business plan is important, you will find that your personality, or those of others in the plot, will account for at least half of the decision.

Use accountants, by all means (you will need one on board for the business plan), but never, ever let them run your business, unless they've either been there themselves or have gone to business school. The problem is, their training makes them very narrowly focussed, and they often fail to see the big picture. Mind you, pilots running businesses have limitations, too, since they're programmed to fly and don't always realise you can make more money by not flying sometimes.

BUSINESS PLAN

This is needed to raise capital - it is a brief sketch of your proposals, detailing how you mean to repay the money, together with how things will be run (this includes the management team). Like a resume, it should be short and to the point, somewhere between a quarter and a half inch thick and, if it is well thought-out, need not be too polished, though it should still look neat, tidy and professional. Here are some suggested headings:

- *Introductory Letter.* Why are you writing the plan (for expansion? An aircraft?)
- *Title Page.* Name phone, date completed
- Table Of Contents
- *Summary* of proposed venture. Description of business
- *How much needed,* and how it will be used
- *Amount already invested;* equipment, market testing, etc.
- *Security* - property, stock and other assets
- *Background Information* - Limited company? plc? Sole Proprietorship?
- *Mission Statement.* Key activities
- *The industry.* Niche available. Competition
- *Management* - organisation, resumes, proposed benefits
- *Target markets* - individuals or corporate
- Advertising
- *Land,* buildings and equipment

- *Operations* - work flow, personnel
- References
- *Financial Plan* - capital requirements and sources to data, profit/Loss-cash flow for at least one year; projected income for at least one year; proforma balance sheet, showing projected current and fixed assets, and liabilities. Break-even calculations
- *Risks and problems;* worst case - what if demand falls, or you get more competitors, or overheads. What's left if you don't succeed? Avoidance of risks. Impact of risks you can't avoid

The magic figure to survive in the small charter world is 500 hours per aircraft per year - that's revenue hours, ten a week, which doesn't include training, etc. Remember, the object of the company is not to fly, but to make a profit so that you can live, or to provide the investor with a return on his capital (not profit necessarily, although sometimes they can coincide). Far too many people forget this, set themselves up with no research, don't market their product, then expect the world to beat a path to their door because they have an aircraft. Even if work does come, more often than not by accident, the same people undercut everybody else around, thinking to put the opposition out of business then put the prices up again. Unfortunately, it doesn't happen like that - they're the ones to go out of business first because they have no cash flow, left with debts, wondering what went wrong. In fact, too many companies operate on cashflow only, and don't put money by for the future, which is how they can undercut everyone else around. The trouble is that they have to guarantee getting the next work, which means undercutting even more into a vicious circle. There is a lot of revenue in flying helicopters - *please* don't make the mistake of thinking that it's all profit and go spending it!

A lot of Aviation Companies owe their existence to a larger parent company that bought an aircraft as a way of spending excess money that can't otherwise be used (otherwise known as a tax dodge), but it's not impossible to survive purely on Aviation without assistance from a Big Brother. Whether it is or not depends on the existence of competition, how big you expect your Company to be and the availability of the work itself (if there's no competition, there might not be any work!)

PURCHASING AN AIRCRAFT

Whatever happens, you will have to get your hands on a machine. Expensive stuff like that (while being an asset in itself) will create massive debt which will require servicing, which in turn means interest. Which company will take care of that? If the Aviation Company itself buys the machine then it will have that much more to worry about.

What is more likely is that an outside company will own the aircraft and lease it to your Company, giving the additional benefit of the equipment being one step removed in case of disasters. Outside aviation it's common practice to place all valuable assets into a holding company that trades only with associated ones, thus insulating them from unplanned contingencies (try looking at a Pure Trust). Where Aviation is concerned, it also legally separates the registered owner from the user. The leasing cost to you will be a total of maintenance costs, spares or engine replacement costs, insurance costs plus a bit on top for contingencies (the spares or engine replacement costs are like depreciation, an accountant's way of establishing a fund for future replacement of machinery).

One creative solution is to raise a loan on something else, say, a house, and use that money to buy the aircraft - you will then get lower interest rates and longer terms.

The aircraft doesn't just cost, say two million pounds. It will also cost what you can't do with the money having spent it - what economists call the opportunity cost. In other words, you lose the opportunity to do something else with it, even if only to sit in a bank account and gain interest. Sometimes it's better not to buy outright but to do it on a mortgage and let the interest gained from whatever you do with the rest pay the interest on the mortgage. With a little shopping around for interest rates (abroad as well) this is entirely possible.

Imagine you have the choice of two aircraft - one relatively expensive to buy, but cheap to operate and the other cheap to buy, but expensive to run. Both do the job you want - well, near enough, anyway. The difference in purchase price between the two may well, if placed on deposit somewhere, more than pay for the increased running costs if you buy the cheaper one. However, this may be low on the list of priorities, as often the purchasing of an aircraft will tend to be a personal decision on behalf of the Chairman.

So, when evaluating an aircraft, first establish what you want it to do - in many cases, a simpler, cheaper aircraft will suit. For example, if you want a helicopter for corporate transport, use a 206B-III, but for training or pleasure flying, a 206A would not only be cheaper, but more efficient, as its C18 engine is not cycle-limited.

Like with a car, look beneath the shiny paint. There's nothing wrong with sprucing something up for sale, but make a thorough examination anyway. Do not do what

one buyer of my acquaintance did - took a helicopter away to lunch, leaving the engineer that he'd taken along (at great expense) alone to look at the books which were written in German! Yes, he bought the wrong aircraft; and deserved it! It sure looked nice, though. Take time to talk to pilots and engineers who actually work with the type of machine you're after - you may find that what you're looking at is OK until the turbocharger goes, which then takes at least three days to repair because it's hidden behind the engine which has to come out completely. On the other hand, another ship could have similar work done in less than half a day and doesn't go wrong in the first place because the turbocharger is not in such a stupid place. Similarly, a particular helicopter could be cheaper to run on paper, but its shorter range on full tanks means that you're paying out for landing fees and dead flying more often, thereby bringing the total operating cost nearly equal to something more comfortable with more endurance (accountants don't often see the big picture).

You need to check the data for rotor or engine Times Between Overhaul (TBO), the Mean Times Between Failures (MTBF) on avionics equipment, amongst other things. Certainly, buy from a company that can provide support, particularly an engineering-based one, and have an independent survey by a competent engineer.

AIRCRAFT VALUATION

Actually, when it was built is largely irrelevant; what counts is the time remaining on its components, since they must all be inspected and replaced at specified times. Equally important is documentation supporting it - it can take longer to verify paperwork than physically survey the aircraft. In this respect, be especially careful when buying from the USA. There are many apparently "cheap" aircraft available, mostly confiscated from smugglers or drug dealers - with no acceptable documentation, their only value is scrap. Also, the regulations for privately owned or agricultural aircraft are less stringent than in Europe, and you may need expensive engineering and/or major components replaced before they will get a C of A. So:

There is no such thing as a cheap helicopter!

Which applies to maintenance as well; if you save money one year, expect to spend it the next. Remember that as well as shipping charges, you may have local taxes and costs of dismantling, packing and erecting when you get it to wherever you are. Shipping is normally All Charges Forward and you will cover insurance.

DEPRECIATION

Because inspection, overhaul and replacement can ensure that a 10-year old aircraft can be as efficient as a new machine, you can't apply this in the same way as a car, or other industrial machinery. However, accountants like it, and it's useful for calculating operating costs, so take the purchase price and give it a one-third residual value, then write down the difference over 10 years. Market prices, though, may vary this. If you have a new engine, or something equally expensive, the machine's value could exceed the new cost.

DIRECT OPERATING COSTS

All manufacturers publish estimates of the cost of operating their machines, in amounts per flying hour, which naturally bear very little relation to reality.

- *Fuel & Oil.* The shorter the sector length, the higher the average fuel consumption. Fuel flow for budgetary purposes tends to be within 5% of max cruise fuel flow for fixed wing aircraft and 10% for helicopters. For oil, only bother with consumption; scheduled oil changes come under maintenance.

- *Scheduled Maintenance* (normal inspection cycles, divided by the hours between them). This is sometimes the second largest direct operating cost after fuel and varies with the flying. Much depends on the cycles incurred by an airframe or engine rather than the hours flown. Wear and tear is felt primarily on takeoffs and landings, when engines run at high power settings and landing-gear and flaps are cycled. Some jet engines (particularly helicopter ones) are therefore restricted to the number of start cycles in addition to flying hours, because of the enormous spread of temperatures incurred in the start sequence. Thus, if your aircraft does relatively short trips your maintenance costs per hour will increase as a result. Whether maintenance is major or minor usually depends on the cost of the item, and the cut-off point is usually left to Company discretion. For avionics, 4% of retail price should reserved for maintenance (assuming 400 hours per year). Add ½% for every 200 additional hours flown, or subtract if need be.

- *Unscheduled Maintenance*, like snags and defects, special inspections required by manufactures and authorities. Add 25% of scheduled for a 206, and 50% for an AStar.

- *Aircraft Overhaul* - removing, overhauling, and replacing components at proper intervals, divided by the interval hours.

- *Engine Overhaul* - as above.

- *Lifed Items* - replacing when time expired, divided by Life Hours.

- *Engine Lifed Items* - as above.

- Labour Rates.

FIXED COSTS

These cost the same per year regardless of the work, so the more you do, the less per hour they become.

- *Purchase/debt servicing*, described more fully elsewhere.

- *Trade in/Resale.* This (hopefully) is a plus item, just resale income for part-exchange. However, the real value of this will be degraded by inflation.

- Depreciation.

- Insurance.

- *Salaries.* If you're an owner-pilot, you could ignore this, but you may have management, operations, mechanics and pilots to pay.

- *Storage.* Outdoor storage is less expensive than hangarage, but storms and damage do happen, as do vandals. The average annual costs of hangarage to be allowed should be roughly one per cent of the equipped price of the new cost of your aircraft.

- *Training*, such as anti-terrorist training, or Dangerous Goods.

- *Services.* Manuals, trade subscriptions, airways manuals and maps. Association Memberships.

- *Tax.* Best left to the experts, this, but there are ways of obtaining an asset on a mortgage offsetting Corporation Tax.

A corporate machine will average 200-300 hours per year, so the total of Fixed Costs divided by that will give you an amount per hour.

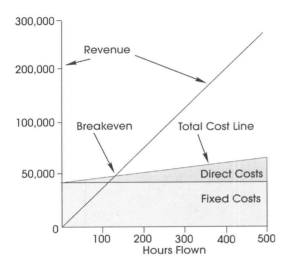

The picture above shows a possible graphical solution to finding the hourly cost of operating an aircraft or two.

INCIDENTAL COSTS

Hotels, taxis, landing, hangarage, handling, tips, oxygen, uniforms, cleaning, office costs (computers, telephone, fax) and freelancers. 25% of the sum of Direct Operating Costs, Fixed Costs and Crew Costs is normally added on top as a contribution to running the business. The total will give the Cost Per Flying Hour. Divide that by seats available to get the Cost Per Seat. Further dividing Seat Cost by the speed will give you the Cost per Seat/Mile.

Obtaining An AOC

Now if you still think you'll be some sort of success, you will need your Air Operator's Certificate, which is required by all operators engaged in Commercial Air Transport. It's applied for on a form together with a CLM (Cheque for Lots of Money) made payable to the Authorities, which will cover all types you wish to include initially. However, subsequent additions will cost the same again, so if you know you'll be adding a new machine later, it makes sense to try and include it from the start. Unfortunately, this causes its own trouble in the form of additions to the Ops Manual and further training costs, as pilots must be qualified on the new type as well. Beware of having too many types in different performance groups.

You can only hold one AOC in one State, and your office and aircraft must also be registered there.

The application form itself is quite easy to fill out. If the Operator is an incorporated body, you will need to know the directors' names, addresses and nationalities, and, if not, the same information with regard to the partners. If there is a trading name separate from the Company name, that will need to be given as well. This bit is quite important, because the AOC is issued to the parent organisation trading as whatever they care to call themselves. The Authority will want to know exactly which trading names are to be adopted.

Otherwise, the only other thing that may need a bit of research is the Maintenance Schedule reference for each aircraft that you propose to use.

The application form and the fee should be sent to the Authorities at least six weeks before operations are planned to commence. Together with all that, you will need a copy of your proposed Operations Manual.........

THE OPERATIONS MANUAL (AGAIN)

Although you don't have to send this with the application form, things will happen considerably quicker if you do, because it needs about six weeks to read it (and about three reviews to iron out the bugs). If it's ready when you apply, some parts of the aforementioned form need not be filled in; you can just refer them to the Manual.

Production of the Operations Manual, which is your way of indicating to the CAA how you intend to operate, is (to use the CAA's words) 'an onerous task'. The quick way is to buy one ready made (from me, if you have my phone number), but there is a pitfall in that, just because a manual has been approved once, there is no guarantee it will be so again.

This is because each company is assigned a different Inspector who will have risen through the Industry in his own way, having different experiences to fall back on. His job is to advise you in the light of that experience (more than being a 'policeman', although that is another function) and assist in the formation of the Company. What one Inspector thinks is OK is not necessarily what another will accept (they use that word rather than "approve", as the latter has legal implications). A typical Inspector will have several companies under his wing, and will therefore have to guard against giving away confidential information. He will normally be the only routine contact a company will have with the Authorities, and his main function when you're up and running is to inspect, report and make recommendations on your performance. On routine visits to the Company, he is empowered to examine any documents or records which must be kept (by law), discussing and resolving any problems that may have arisen during your operations. Your AOC is reviewed annually on the basis of his reports and is non-expiring, provided that the annual charges are paid up to date (based on the throughput of traffic that a Company has) and you keep your nose clean.

The Authorities require a copy of your Ops Manual for its own records and for instant reference in case of queries. As it is the primary indication (to them) of your operating standards, it makes sense to produce the Manual in the best possible way. This is psychological - if they see a well-presented manual on the shelves, then they're likely to be more convinced that the rest of the Company is likewise (well, wouldn't you?). So you are doing yourself down if you skimp on the Ops Manual, no matter how boring it may be to produce it, but that's been discussed already.

BACK TO THE AOC

Having submitted the application form and the manual, you sit and wait for a response, during which time the Chief Inspector allocates whichever one he thinks will suit you. While the Manual is being read you can get your pilots checked out by your Training Captains and the system streamlined for the proving ride. For setting out the office, etc., see *Running Things*, below.

Your Inspector meanwhile will visit your proposed offices to ensure that they meet certain requirements (such as the Chief Pilot having his own office and being able to see the aircraft operate, the numbers of clerical staff and machinery relative to Management). They not be on an airport, but being away from one does cause problems, certainly for keeping track of fuel states (so you can calculate your payload instantly if you get a quick charter). Hopefully by then he will have produced some proposed amendments to the Manual, but it could actually be read through for the first time in front of you, just before lunch on the first visit. When he is happy with that, and your offices, he will want a proving ride (with a line pilot) on your aircraft. The ride itself is not a check of the pilot's ability, at least not in the sense of a Proficiency Check, but more a check of the Company procedures, which is why it should be done with a line pilot, to see if the system works. It's meant to be a simulation of a complete line operation and will be about an hour or so long. The Inspector will pretend to be a passenger and will expect to be weighed, briefed and otherwise treated exactly as per the Operations Manual. Almost the first thing he will make a beeline for on arrival at the office is the ANO to see if it's up to date! The same goes for maps and other documents. You don't need full copies of the Air Pilot or NOTAMs, provided you can prove you have adequate access to any flight planning information you may need, including weather.

After the ride, assuming all is well, the AOC should be granted in due course, possibly after a few more changes to the Manual. The issue of the Certificate signifies only that you are considered 'competent to secure the safe operation' of your aircraft - it doesn't relieve you from any other legal responsibilities that you may have, whatever they are.

Once you have your certificate, your Inspector will pop round within a month and thereafter about every six months or so to ensure the continued competence of the Company, including any outstations or agents that you may employ.

Running Things

Aircraft must achieve high utilisation for maximum cost-effectiveness - an aircraft on the ground is not earning money. While corporate flight departments do not make a profit as such, the comments here apply equally to them, as efficiency still helps the bottom line. Also, if you are operating an aircraft by yourself and have no Commercial Air Transport experience to fall back on, you ought to realise that a good office environment back at base is very important to the overall operation. The following pages will give you some idea of what's required to run things properly, with a little information on the corporate scene that should be read as well because it's all relevant (and it saves me typing it twice).

The various functions to be filled include planning, day-to-day operation and administration. The bigger the company, the larger the departments handling these will be. You may find you need none of these, but it's still worth knowing what they get up to. Planning covers everything from long term management decisions through scheduling flights, minimising dead flying and taking care of maintenance, although some of this could be regarded as day-to-day operations. Administration is the only part likely to be really separate, but even here there is likely to be a lot of blurring between departments as staff wear several hats.

In practice, you will find all the above activities (with the exception of Top Management matters) more than adequately looked after by the Operations Department in the average small company, and this is what we will mainly be dealing with in this chapter.

THE OPERATIONS DEPARTMENT

Operations is in immediate control of all day-to-day business, the focal point of its activities being the Ops room where, depending on the extent of your activities, will be found the Operations Staff, secretaries and the rest (if they can all fit in).

The role of the Ops Room (including staff) is to ensure that the right aircraft is in the right place at the right time and that everyone concerned is aware of what is happening, being pre-warned of any problems which may be expected. If this cannot be achieved for any reason, Operations must initiate remedial action and minimise inconvenience to passengers, who are (after all) the source of the Company's income.

One of the ways Operations keep track of events is with Movements Boards (boards are quite useful, and you will find that several will be required for AOC operations, including Pilot Qualifications, NOTAMs, and everything else you may think of). These boards should be kept well away from prying eyes who may pinch your business if they see destinations and customer names, so don't put them near windows. Movements Boards should be constantly updated as they're a major reference point. What goes on them is up to you - just use whatever information you think will be needed. The biggest Movement Board of the lot is the map, which will usually have a string-and-weight arrangement with a Nav Ruler that makes it easy to calculate complicated distances.

Linked to Movement Boards is the Diary. There will be a scruffy one that's used daily, but there should also be a backup filled in after the day's work. In it should go all the scheduled work, upcoming pilot and aircraft checks (a week or so before they're due).

As mentioned before, there will be a quotes file. It is suggested that this be loose leaf, each page being filled in at the time of each query. If a trip looks like it's going to happen, then that page should be put into a pending file until confirmed, when it's put into a Diary file.

The Diary file is simply 32 file holders, not necessarily in one book, representing each day of the month plus one, and all prepared documentation for a flight is placed in the file for the relevant day. The benefit of this system is that information that's only valid for a day can be put in there as well, which makes it easy to bring it to the attention of the staff concerned.

You can see that communications are beginning to be of vital importance - a good communications network is an essential part of modern aviation. Without knowing as quickly as possible what's going on, it's very difficult to plan ahead and foresee problems. Many methods are used, including satellite phones, and you should encourage your pilots to use it often, because it helps with scheduling if you know where they are, aside from allowing quick responses to incidents.

The most common, however, is the telephone, and the correct use of it saves many problems. The first problem is that there is no record of what's been said, so important messages and decisions made on the telephone must be followed up immediately by hard copy. When taking down a message, always ensure you have the correct information and names, so you know who to blame later. The telephone should be answered as soon as possible, and before answering, be sure you have a fair chance of helping the caller. When answered, they should not be left holding. If they have to be for any reason, ensure that nothing can be overheard that shouldn't be! A definite reason linked to holding is essential, and regular assurance that the problem is being dealt with is helpful as well.

Don't use jargon or be familiar with people you do not know; refer again to previous comments on being an ambassador of the Company. Always terminate a phone call leaving a positive impression.

CORPORATE FLIGHT DEPARTMENTS

Here you may well find yourself actually in charge of a Flight Department in the proper sense of the word. A charter company in fact, without the necessity of bothering with charges, although if they are offset between companies within the same group (known as *chargebacks*), they will normally be handled by Accounts.

Chargebacks are one way of allocating time between users in large companies, paying for the machine on paper, but if the rates are too high, the end result is that the departments who need it most can't afford it and therefore can't use it, which seems a bit pointless. A side-effect is that it opens the door to small charter operators who can do the job cheaper, and then money flows outside the Company instead of staying in it. Another is that Accounts have a chance to do a bit of empire-building, as they are the only ones who get any work out of it, namely chasing money round in a circle.

If you're employed as a full-time pilot, your Company will probably already have an aircraft, so it's unlikely that they will charter in except for times when they've lent their aircraft out. Leasing (self-drive) is a good half-way house between chartering and owning. The cut-off point where owning an aircraft makes more financial sense than leasing is about 200-250 hours a year, so the average flying rate for Corporate aircraft is at least 200-300 hours per year, but some get up to 600 or 700.

Your company may do things the other way round and lease their aircraft out to commercial operators. This causes problems, especially where allocating priorities are concerned, and if this is done extensively, Management will have to get used to the idea of either going without their aircraft or hiring another. Sometimes trying for extra money on the side defeats the object of getting an aircraft in the first place, but that's not your problem.

Despite chargebacks, Company policy may dictate that the costs of operating the aircraft are not actually charged against the Flight Department (for instance, in one or two companies they come under Sales). Whoever looks after it, Management will (naturally) want to know where the money is going and how much will be wanted next year so they can budget properly. This means getting involved in statistics because as a Company employee you're being paid anyway and you have nothing else to do, right?

If you do get lumbered with all that, you may find it easiest to add some columns to your Tech Log to fill in as you go along. Who is flying is often more important than just noting how many, so possibly you may like to account for Corporate, Divisional or Production employees. Don't forget marketing, freight and sales. Some useful tools for making Impressive Reports (especially in comparison with other forms of transport) include:

- *Average speed.* The number of miles flown divided by the number of hours airborne.

- *Average fuel consumption.* Fuel burned divided by hours airborne (chock to chock), including ground runs, etc.

- *Load factor.* The total number of passengers divided by the total number of sectors, a sector in this case being a nonstop flight from A to B regardless of the ultimate destination. Non-revenue or non-productive sectors (positioning, training, air tests) if included will adversely affect the load factor, but will give a better picture of operational efficiency.

- *Ac Miles travelled.* Total miles travelled by the aircraft.

- *Pax miles travelled.* Aircraft miles travelled multiplied by the Load factor (if you're wondering why these need to be calculated, it gives a quick indication of productivity).

- *Cost per hour.* Total cost divided by the hours flown.

- *Cost per aircraft-mile.* Total cost divided by the number of miles flown.

- *Cost per passenger-mile.* Total cost divided by passenger miles (obtained by dividing aircraft miles by the load factor).

BEING A CHIEF PILOT

First of all, you should not be on the Duty Roster. The job requires so much management that you become ineffective on both sides if you try and do too much - there's simply not enough time (you can see the responsibilities above). Any flying you do should be strictly to keep current and step in when there's a shortage. As mentioned before, you will be the main point of reference for officialdom, because you dictate the flying policy of the Company. Another plus point is that you get to argue with Management when they want to put commercial pressure on. Very often you will find yourself in a position where you're made to seem very unreasonable:

> *"Oh, go on, it's just a little snag. You've got another one and if you don't use this aeroplane I've got to get one all the way from up North. So what if none of the gauges work?"*

or (and this is a true one):

> *"It's only a small crack in the fuel tank, and it's at the top, so if you don't fill it all the way up, it won't leak out."*

are common conversations between a Managing Director and a Chief Pilot. Come to think of it, it's more common between MD and pilot so the Chief Pilot doesn't know anything about it. If a pilot comes to you with a problem like that, you must back him to the hilt, even if you think he's wrong. You can always sort that out in the pub later.

All this involves your personal integrity and credibility - referring to previous comments about money management, very often if you need to lease a plane, the lessor will take your word for it (as holding a position of responsibility) that they will get paid. In fact, less-than-honest Management have been known to hide behind the reputation of members of their staff, with the resulting loss of several peoples' good names when the money wasn't forthcoming. You have been warned!

BEING A BOSS

Money is not a motivator. If people are unhappy, they will be just the same very soon after a pay rise - watch what happens to your engineers if you don't give them enough spare parts. You will find that, left to themselves, most people do the job well enough without supervision, and want to do well at what they do. You will get the best out of your staff if their goals align with those of your Company, and that happens when you give them the ability to make decisions, or assume responsibilities. Any problems are 99% down to bad leadership. You're not as concerned with what people get up to when you're there, but what they do when you're not there!

QUALIFICATIONS

These are sometimes dictated by the insurance company, customers (like oil companies with their own safety departments), or Personnel, who mostly don't have a clue.

LICENCES

There are generally two grades of licence that entitle you to earn money as a pilot, the *Commercial Pilot's Licence* (CPL) and the *Airline Transport Pilot's Licence* (ATPL).

Note: If you are not going for the ATPL, and in some countries, even if you do, you need to budget for additional qualifications, since you are virtually unemployable with just a CPL(H). You either need more hours (minimum 500), a mountain course, an Instrument Rating or an Instructor course. Note that instructors are not well paid!

The CPL(H)

The Commercial Pilot's Licence (Helicopter) covers you for command of light helicopters, say, up to 5700 kg. You can't have one until you reach a certain age, but you can take the exams beforehand - sometimes up to a year ahead. Your flight training will include the usual mix of night, instrument, dual, cross-country, etc., and ground training will cover Principles of Flight, Air Law, Meteorology and some technical stuff, and you may get credit for previous experience, such as aeroplanes, military wings, and the like (because regulations change, check the details elsewhere).

ATPL(H)

You get an Airline Transport Pilot's Licence when you are much older and get a lot more hours under your belt (1000 in Europe). It covers you for command on anything, subject to type rating. In some countries, you can just get the hours and take the exams (USA). In others, you also need a check ride in a helicopter that requires a co-pilot (Canada). In Europe, on top of the type rating for a big machine, you also have to do 500 hours in a multi-crew environment (or a course), so an ATPL is virtually unobtainable outside a company.

EASA

The process for European licences is not as straightforward as it is in, say, Canada or the USA. In fact, the EASA exam setup is currently an international joke, for which the perpetrators should hang their collective heads in shame (however this will change in around 2018).

Let's take a look at the syllabus first. About 40% of it would, in any other country, be used for a Flight Engineer's licence. In other words, it is overly technical, and there is a lot of it. Possibly the reason is that the exams were originally designed to form part of a degree course and, as with most degrees, consists of material that is there for credibility rather than anything else. As a result it is quite possible for a potential pilot to get an exam paper full of irrelevant questions and not get examined on anything useful at all.

Which leads me to the questions - as they match the syllabus, more or less the same proportion has the same status, i.e. around 40% are irrelevant for pilots. What's left is riddled with bad English, bad punctuation, misspellings, multiple correct answers, multiple wrong answers (20% of the total database), some in the middle - in short, it's a major SNAFU.

The result is that is it not possible to pass the exams on knowledge alone, as you might expect if you are a seasoned professional with an ATPL(H) from outside Europe, as you are entitled to do if you have over 1000 hours multi crew.

The ATPL(H) has 14 subjects, and the CPL(H) has 13. The exams should be passed within 18 months of sitting the first one. You can have up to 6 sittings in that time, but only up to 4 attempts at any subject, otherwise you must resit them all, after further training. A pass in all subjects for the ATPL is valid for 7 years after the date of expiry of the type rating. The IR element is valid for 36 months.

Flying Training

As long as you meet the requirements, your school's Head of Training will decide how much flying training you need before the test. Most schools will probably expect you to do at least 10 hours to ensure that all manoeuvres are covered, or, rather, that you are comfortable with them.

For more information, refer to **www.captonline.com**.

CANADA

The CPL needs at least 100 hours in helicopters, but with a PPL(H), Canadian or otherwise, you get away with 60. You must also do at least 40 hours ground school, but if you don't already hold a PPL(H), you must do another 40, for a total of 80 (this is expected to increase at some stage). You can take the exams any time at any Transport Canada office. If you don't manage the night hours, you can still get a licence, but for day only, if you meet the dual and solo times. The restriction will be removed once you get the time in.

For the ATPL, you must have at least 1000 hours, 600 hours in helicopters (instrument ground time does not count), pass two more written exams and take a check ride in a helicopter that needs a co-pilot. The catch is that there are only a few helicopters that always need two pilots, the Sikorsky S61/92, the AW 139 or the Super Puma, though you could also count the S-76 or 212 or similar when IFR - apparently, as long as the flight manual carries the magic words "requires two pilots" somewhere inside, that will do.

The written tests are quite practical, with multiple-choice questions which can be easily done inside 3.5 hours, and all you do is turn up at any Transport Canada Licensing Office (one in every city) during business hours, with a Letter of Recommendation, proof of at least 100 hours flight time, made up of the usual night, cross country, etc., Canadian medical certificate and the fees. Compared to the three days or so required in UK, this seems ridiculously short, but all is not as it seems - you've still got the flight test!

Before taking off, you and your examiner retire to a quiet room for an intense session of question-and-answer, with a fair emphasis on actually being a commercial pilot. For example, you could be asked for your reaction to a passenger who turns up with far too much baggage and insists that he got it all in last week, as if being interviewed for a job in your examiner's own company. Otherwise, you must know what documents an airworthy aircraft requires, their validity and when they must be carried, how weather

is reported, what types of airspace you might fly through on your trip, who you would call to pass through them (many aerodromes are operated remotely), danger areas, etc. You will not be expected to know map symbols and the like off by heart, but you must know where to find the information. During the preflight briefing, the Flight Manual and Canada Flight Supplement (Pooley's on steroids) will be used extensively.

After a couple of hours, you will move on to preflight the aircraft, and brief the "passenger", who is obviously the examiner in disguise, and able to change altimeter settings, hold the controls while you sort maps out, etc. Although you will have planned for a 2 hour flight, after about 20 minutes, when it's obvious you're going in the right direction, and you've come up with a reasonable groundspeed and amended ETA for the destination, the trip will be aborted and you will go into the other stuff, maybe landing in a clearing, some instrument flying with timed turns and unusual attitudes, electrical failures, etc. During this time the examiner will be looking for good cockpit management and liaison with ATC.

There are certain peculiarities to the industry in Canada that make it interesting to someone from Europe, but the remarks below apply to other very large countries with few roads.

One of the first things you will have to get used to is distance - it's quite possible, for example, to fly the equivalent distance from London to Manchester (that is, halfway up the country) and not see a soul. It's for this reason that companies operating in such areas have fuel caches, and a pump and water detection kit which *always* stays with the machine, despite how tempting it is to make room for baggage. These caches are refilled at regular intervals and it's part of a pilot's responsibility to report the contents back to base on the regular daily check, which is typically done in the late afternoon. Actually, position reports are not done much these days - most companies use satellite tracking which is automatic and requires no talking. The Ops Manager must know where you are for legal reasons, but the daily report (if done) is also for scheduling purposes, namely to tell you what you're doing the next day (and you thought you were going home!) It's generally OK to use fuel from another company's cache, but it's considered good manners not to do it too often, or to use too much, and to leave a note somewhere as to how much you've taken and to let them know about it as soon as possible, and replace it.

In many cases, you will find that some customers have as much of an idea how to operate a helicopter as you do, if not more, and will have a considerable input into the type of work you do. Aside from the usual stuff on the oil patch, or fire suppression, Heliskiing is popular, too, with lots of short trips in the mountains, and the added attraction of trying to land on postage stamps and what look like very rudimentary landing stages made out of logs, which can only be approached from one direction, regardless of the wind. It doesn't help to look down!

So, the work is definitely not boring, but an extra twist is doing it all by yourself, which is not actually for commercial reasons - you will be typically taking equipment into places that ground crews can't get to anyway, and, in the Arctic, you can't risk leaving a team on the ground in case you don't get back. Other aspects also come into play - for example, you need a police certificate for firearms, because you may need one if you find a bear.

European pilots (well, military ones, at least) have it drummed into them throughout their training that they shouldn't fly over trees with one engine. However, in many areas, all you can see for miles around is - trees! Under these circumstances you rapidly learn to trust your engine, and be nice to the engineers who look after it. You also learn to take extra care over preflights, and start doing afterflights more often, where oil leaks are more obvious, and there won't be time to fix things in the morning. You will also be taught elementary servicing tasks.

Duty Hours are longer, too. The average day is 14 hours, in which you can cram as much flying in as you can within the bounds of reason. You can also be on duty for up to 42 days, or 60-plus on a fire.

A season flying helicopters in Canada is highly recommended for sharpening up your flying skills - it will certainly make you a way better pilot.

USA

Essentially similar to Canada, except you need more hours (150) for a CPL, 50 of which must be on helicopters (the rest on powered aircraft). You also need 100 hours PIC, with 35 in helicopters and 10 cross country, plus 20 hours on the subjects in FARs 61.127(b)(3), to include a 2-hour cross-country flight by day and night in VFR at least 50 miles long, plus three hours' practice for the test in the 60 days before taking it, on top of 10 hours solo with a cross country flight having three landing points and a segment over 50 miles long and 5 hours in night VFR with 10 takeoffs and landings with a circuit in between.

The ATPL can be obtained on any aircraft. You need 1200 hours, with over 500 cross country and 100 at night (15 in helicopters). 200 hours of the total must be in helicopters, with over 75 PIC, or P1/US (or any combination). You must also have 75 hours on instruments (actual or simulated), 50 in flight, and over 25 in helicopters as PIC or P1/US (or a combination). You can only claim up to 25 hours in a simulator, or 50 on an approved course.

Oh yes, you must also be of good character!

Civil Aviation Authority

European Union
United Kingdom Civil Aviation Authority

APPROVED TRAINING ORGANISATION CERTIFICATE

GBR.ATO-0129

Pursuant to Commission Regulation (EU) No 1178/2011 and subject to the conditions specified below, the UK Civil Aviation Authority hereby certifies

CALEDONIAN ADVANCED PILOT TRAINING LIMITED

Wycombe Air Centre Building Wycombe Air Park SL7 3DP	2a Courtwick Road Littlehampton BN17 7NE	C/o Helicentre Aviation Limited Leicester Airport Gartree Road Leicester LE2 2FG
	C/o HeliCentre B.V Arendweg 33 8218 PE Lelystad Netherlands	C/o Cloud 9 Helicopters LLC 11610 Aviation Road West Palm Beach Florida 33412 USA

as an Approved Training Organisation with the privilege to provide Part-FCL training courses, including the use of FSTDs, as listed in the attached course approval.

CONDITIONS:

1. This certificate is limited to the privileges and the scope of providing the training courses, including the use of FSTDs, as listed in the attached training course approval.

2. This certificate is valid whilst the approved organisation remains in compliance with Part-ORA, Part-FCL and other applicable regulations.

3. Subject to compliance with the foregoing conditions, this certificate shall remain valid unless the certificate has been surrendered, superseded, limited, suspended or revoked.

Date of issue: 15 February 2013
Date of re issue: 30 April 2015

Signed
For the UK Civil Aviation Authority

EASA FORM 143 Issue 1 – page 1/3

SPECIALISED TASKS

Some of the more exotic things you can do with helicopters include wildlife capture, bombing avalanches, rapattack (that is, dropping off people to fight forest fires without landing, otherwise called dope-on-a-rope), aerial ignition, water sampling, where you hover over a body of water and a scientist dips the equivalent of a jamjar into it (I still haven't found out why they don't just use a boat), or frost control, where a large barrel of oil is lit to provide smoke that will indicate the level of an inversion. You then fly with your rotors just above the smoke to bring the warm air down and prevent frost on crops.

Slinging, Mountain Flying, SAR and Offshore are covered in separate chapters.

Note: Some of this is *dangerous*! Don't try it without training! The fatal accident rate in 2004 was highest with helicopter seismic operations by a wide margin!

AERIAL APPLICATION

Aerial application (of pesticides or fertilizers) means either crop spraying or top dressing, the latter being used in forestry (although you can spray cut blocks with booms on). Top Dressing is more akin to load slinging, except you use engine driven devices like buckets to spread solutions over forests. Unlike crop spraying, it can be done in strongish wind conditions, but, otherwise, it's characterised by always being in, or very near, the avoid curve and many other situations that you're taught to avoid normally. You can tell with forests that have been sprayed in the early stages of their growth as to whether the pilots were successful or not - you often see trees shorter than others, which is where they missed.

Helicopters are particularly useful when leaves need to be sprayed underneath, due to the downwash. At low speed, you can spray a small area underneath the flight path. At higher speeds, the wake helps spread the load behind and to either side (the term rotor wake means all the air displaced by the helicopter, as opposed to just the downwash). Knowledge of wake management will therefore help you become a better spray pilot.

The Rotor Wake

This changes within three distinct speed ranges:

- *Up to 20 mph*, the air moves primarily downwards, most of it descending from the outer edges of the blades, so you get a relatively calm area around the fuselage (in other words, you are in the middle of a ring, like a doughnut - you can see this by hovering over water). The force in the outer ring can agitate foliage so it collects chemical above and below, in a fog over a relatively large area. Slow speeds, however, cost money.

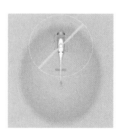

- *At 18-22 mph* (on a Bell), the annular ring shortens in the direction of movement to become an ellipse. Above 20 mph, the annular ring disappears, and a large amount of separate, small airflows coalesce to provide an area of ill-defined airflow with a general downward direction.

- *Above 35 mph*, two distinct rotating vortices are formed from directly behind the machine to a long way behind, assuming no outside influence (they are fully developed about 1 rotor diameter behind the mast, and can be sustained for up to 2500 feet). Each vortex starts from where the annular ring would be in the hover, and is relatively calm in the centre (the centre-to-centre distance between them is just under the rotor diameter, and slightly displaced from the centre towards the retreating blade). Regard them as large funnels extending rearward and downward, getting bigger as they go. The point is, they can direct chemical into foliage. Ground cushion, however, can cause them to separate, because there is nowhere else for them to go. There is still a downward flow as well as that from the vortices.

Photographer unknown

Particle sizes at low heights should be larger to prevent them being sucked into the vortex areas, where they will not be effective. You would spray at higher levels if you want to get the underside of foliage, as with an orchard, and make full use of the vortices. In this case, particle size would be smaller, but large enough to fall out of the air stream. The denser the foliage, the higher the air stream needs to be. It has been found that, between 40-80 mph, the swath width can be relatively constant, meaning that you can get the same physical coverage while slowing down to concentrate on more heavily infested areas. Application is inversely proportional to airspeed, so 2 gallons an acre at 60 mph translates to 3 gallons at 40 or 1.5 at 80.

Note that there is a different optimum particle size according to the foliage - a good reference book for your library in this respect is *Concentrated Spray Equipment, Mixtures and Application Methods*, by S F Potts (Dorland Books, NJ).

The Procedure

Crop spraying, like slinging, is very satisfying when you get a good rhythm and an efficient team that keeps you in the air as much as possible. Unlike it, however, you will be operating a heavy machine with unwieldy spray booms (actually just like bouncy missiles, if you ever did that sort of thing) at some speed in confined areas, and the low level manoeuvres will require a lot of co-ordination and forward thinking. For example, you have to continually

keep a note of the ground so when you turn round you can pick up from where you left off and remember where you've been already. When lining up, you need to get the speed and heading right, and pick a spot to aim for so you can keep straight, all in the space of a few seconds. However, there are GPS-based systems that will even turn the booms on and off automatically. There's usually very little wind to help you, because of the legal restrictions on wind speeds and the possibilities of Spray Drift (see later), and spraying cotton in Australia is actually done at night. In fact, you can only really count on about 5 hours' productive time during the day; 3 in the morning and 2 in the evening. Anything else is a bonus. The big problem with doing it at those times is that the Sun is always in your eyes from one direction, which may not be ideal if there is any wind about.

*After a number of spray booms got damaged by being flown into the crop a memo was placed on Dollar Helicopters' notice board:

"Do not fly at 2 feet when 3 feet will do!"

The plot is to fly between 35-45 knots at about 5 feet* along the "grain" of the crop (if you don't need penetration), at the end of the run pulling up and pivoting around (torque-turning) to face the other direction on the end of the boom that is pointing into wind, so you start where you left off (the speed and height would be about 50 kts and 50 feet in forest blocks, allowing for obstacles).

Thus, you turn into wind at the end of each run, progressing towards the wind direction. The trick when turning downwind is to pull up, turn halfway, then let the machine fall into the right spot, with maybe a little help from some backward cyclic. Hopefully, after each run, you can still see your load settling and can use it as an aiming point for the next one, so you could liken it to using a paintbrush, even to the extent of using pedals to twist the swath into the fiddly bits of forests. Expect to overlap about 10%. The reason for starting from a downwind position is so that you're not flying into your own spray. Before you start, however, make sure you have the right forest blocks, and you have the permits for them!

A typical load takes around minutes to put out, then you go back to the truck for another. Loading takes 30-45 seconds. Typical fuel is enough to carry 5-6 loads. It is said that you can often tell an ag pilot by the way they drag their skids before taking off because they are so used to being overfilled with chemicals.

Pesticides come in various forms, as solid, liquid or gas - in general, *insecticides* kill insects (it makes them so depressed that they commit insecticide - joke), *herbicides* kill plants (i.e. weeds) and *fungicides* make short work of, well, fungi, but even these classifications can be further broken down. With insecticides, a *stomach poison* must actually be ingested, while a *contact poison* needs merely to be touched by the insect. A *systemic insecticide* can be applied to one part of the plant from where it spreads by itself to the rest, although it may need to be applied in a particular way or place, such as at the root or on the leaves.

Herbicides can also be selective, in that they go after a certain species of plant. *Residual herbicides* provide long-term control, sometimes for up to three seasons. However, timing is important - they must be applied when the plants to be protected are strongest and the weeds at their weakest. You might find *pre-plant, pre-emergence* or *post emergence* types, which really speak for themselves.

A *protectant* is a fungicide designed to protect a plant rather than do damage to a fungus. *Eradicants* are used when it's too late. In with the mix may be an *inert substance*, which might be talc in a dust formulation, or a petroleum product to assist emulsification.

Aerial Application, in UK, at least, takes place under a "Certificate", which is broadly comparable to an AOC. Other countries will have a similar system. Under its terms a Ground Operations Manager must be present, with some minimum qualifications. There should also be a field support engineer, who monitors aircraft performance and attends to routine servicing (the same person could do both jobs). There may also be a flagman, who marks out the areas to be sprayed and the routes to be followed.

Other ground staff include loaders, who mix and load the solution, which usually requires fast action to keep things going but, when things are happening quickly, there is more chance of spillage and contamination.

Wherever you are, though, it is likely that you'll need some ground school in order to pass exams for registration as a spray pilot, which may involve some calculations about nozzle sizes, etc. You must also learn to *read the label on the product*, which will have some legal status. It will contain instructions concerning equipment required, first aid procedures, compatibility, amounts to be used, and where, and storage (pesticides are often Dangerous Goods, as well). If you spray outside the conditions in the label, you will not be legal.

To make money out of spraying, set prices are usually charged for an area, which means the quicker the job is done, the quicker another can be started - the problem is that your company's accounts are based on flying hours, not acreage. The aircraft should be placed centrally, and its production (in acres per hour) will depend on payload available, endurance, dead time between sites, volume of work at each site, terrain, pilot's experience, time spent on the ground reloading, rate of application and weather (I think that's all). Organisation, however, is most important. If you keep changing chemicals, the whole pattern of work will be disrupted, so grouping crops that require the same cover on a regular basis can get rid of not only dead ferrying time, but also unnecessary cleaning of tanks. Very often, you can expect to top up with fuel after every delivery, to get the maximum spread.

To minimise weight, liquid solutions will be more concentrated than normal, and exposing yourself and ground staff to the spray must be regarded as a possibility, either by *ingestion, skin contact* (sometimes through the eyes) or *inhalation*. Remember, whatever it is kills things, so it will be toxic - the larger the *LD50 value*, the less it will be, but below 10 is regarded as extremely toxic (LD50 indicates a dose that will kill 50% of test animals within a certain period, and refers to acute toxicity). Minimise skin contamination with rubber gloves and boots (unlined), and clean clothing which includes long trousers and sleeves, which is just what you need on a hot day, especially with coveralls on top (wear trouser legs outside boots). You don't need use anything waterproof unless you expect to get drenched, or are working in a mist. Cover cuts and abrasions. Flaggers can avoid exposure to whatever is being sprayed by keeping out of the way.

Spray Drift

This is the movement of whatever you're spraying to areas it was not intended for. It's undesirable, not only because it reduces the chemical used on the job, but it also causes damage in non-target areas due to concentrated amounts accumulating downwind, sometimes more than that applied to the target (it's also trespassing). Spray Drift is affected by greater wind velocity at height, volatility of the solution, temperature inversions combined with spray pressure, nozzle spray angle and air movement around the aircraft. You can reduce the chances of it by releasing large droplets close to the target, by:

- Flying as low and slow as possible

- Locating nozzles away from rotor tips

- Placing the spray boom as far forward as possible

- Orienting nozzles backward and spacing for uniform patterns

- Using low nozzle pressures with larger orifices in the nozzles

- Spraying when winds are light and the air is cool, about 4 am

- Using herbicides that do not produce damaging vapours

- Using buffer zones, particularly around water

You can also modify the solution with additives that produce more viscosity (*adjuvants*), or with an *invert emulsion* system, which will apply a mayonnaise-like material. They both have disadvantages, though, in either needing specialised equipment or mixing techniques.

If you're trying to increase the volume of solution over a particular area, you're better off tinkering with the nozzles rather than reducing speed or raising the spray pressure. This will help to avoid small droplets, or 'fines', that are more likely to drift. There are some crops, however, that

require good coverage and small droplets, which is the opposite. This is where you need to be a bit of a chemist as well as a pilot to satisfy the customer.

Vapour drift is a similar effect arising from evaporation.

Accurate records are essential, not only as aids to the business but also for later complaints of drifting. It's helpful to note such items as nozzle size and spacing, wind velocity and weather conditions, rate of application and time on task, together with a diagram (which you should have made anyway, for planning purposes - many pilots use an aerial photograph and keep a log of the tracks flown with a marker).

Seeding

When a pipeline, or similar, has been laid, the countryside has to made to look attractive again as soon as possible. A large hopper full of grass seed is used to do this, and you will get to do a series of runs over the pipeline to spread it:

The hopper is basically a large fibreglass bucket that weighs about 300 lbs. It has a large petrol motor on the bottom, about the size of the average lawnmower engine, with a gate that is opened and closed from the cabin, and the important thing to remember is to positively switch off after a run, otherwise you will drop the seed everywhere. In other words, just letting go of the On button doesn't automatically stop the flow.

The other point is that, like a longline, it is attached to the helicopter in two places, that is, the hook and the electrical cable, and the latter has to be pulled out by ground staff or you will strip the cables when the hook is released.

If you have a lot of ground to cover, think about placing the seed on a truck and operating from various points around the countryside, which will save a lot of dead flying time. Lastly, depending on the machine and how far you have to fly, it's not uncommon to top up with fuel after every delivery.

FIRE SUPPRESSION

Fires (in Canada, at least) generally start from lightning strikes*, or the odd cigarette butt lying around but, in California, the Santa Ana wind is often the culprit. It is a hot wind from the desert, which gets hotter as it descends, and by the time it is ready to do some damage, you are in the equivalent of a fan-assisted oven.

*Lightning can be up to 50,000° in temperature - when it hits a tree, the sap boils and the heat can be held internally for days until fire actually breaks out).

The types of fuel behind a fire can affect its intensity and staying power. They are officially classed as *light, medium* or *heavy,* as with grass, brush or timber, respectively. They are almost never found in isolation, and will feed off each other when a fire starts. Grass gains and loses humidity very quickly, and burns very hot, but very quickly. Brush is more stable when it comes to humidity, but, once lit, will provide a steady source of heat energy for the heavier stuff, such as forests, or even buildings, which require the most effort to put out.

The heat rising from a fire will pull in air from the surrounding area, rather like a thunderstorm does, and on a calm day, the fire won't move a lot. On a windy day, however, not only will there be extra oxygen, but the wind will push the heated air forward, to dry out and ignite whatever is ahead. Spot fires, caused by sparks leaping outside the main fire, are always a possibility, so a good knowledge of what the wind is up to is a good part of your armoury (watch those fronts!) Fires going uphill can also move quickly, and will also cause spots on the lee side.

Fire suppression is a concentrated effort which doesn't stop when the flames are out. In fact, only about 50% of the area inside a fire is actually burnt, although some will be more so than others. Flat terrain will increase a fire's spread. You may be asked to turn your hand to many jobs at the drop of a hat, including:

- help fight the fire, with buckets, attached to the helicopter or on a long line

- move men and materials between camps and fires

- scan with FLIR (Infra Red)

- report fire dimensions back to the fire boss

- recover equipment afterwards

- search for hotspots (with IR)

- Initial Attack patrols

- Command and Control

See also *Hover Emplaning & Deplaning* under *Police Operations.*

Generally, what job you get is determined by the fireboss the night before (once the fire's organisational structure is in place) and you get your orders in the morning when you turn up for work. Of course, what often happens is a system of organised chaos, where it's hard to tell if anyone really knows what's going on and you end up fighting a little bit all by yourself, or even two, as I ended up doing one day. Luckily there was a lake right between them both.

Otherwise, if you're not part of the operation, you should not be within 5 nm and below 3000 feet agl of any fire's limits. Anyone joining in should be in touch with the *bird dog,* if there is one - this is a light plane or helicopter used for air traffic and controlling the water tankers, and you need to be told when they are coming so you can get out of the way for 15 minutes or so. In some provinces, you need the bird dog's permission for any move you make.

If there is no bird dog, the local fire authority should have a flight watch frequency (UHF or FM), and the fire itself will have an aeronautical one. Expect also to need a VHF FM frequency for the ground crews, which means you have to listen out on three very busy frequencies (sometimes, you might carry a fire officer who talks to ground crews for you).

Fires are attacked in three ways:

- The **direct attack**, where water (plus foam or retardant) is dropped directly on the flames with the intention of extinguishing them.

- The **indirect attack**, where drops are made next to the fire to slow it down or reduce its intensity.

- The **combination attack** is a mix of the above, used to protect assets in the line of the fire.

Line building is a line of drops along the flanks of the fire, overlapping previous drops, to pinch off the head of the fire, as it is often too hot to drop directly.

Water bombers will be used either to drop retardant or water, sometimes with foam. Retardant is phosphate fertiliser and water, with a dye so they can see where it hit. It is not actually used on the fire, but around it, so it can be contained in a smaller area and allowed to burn out. Skimmers (like the CL 415 on the left) pick water up from nearby lakes, but others get reloaded from airstrips. There will often be dozers trying to create a break round the fire if it is small enough. Bombers do not usually get below 150 feet, as the water pattern gets disrupted, so your safest height is below that, amongst the smoke, so be careful.

Helicopters also make use of handy sources of water, like swimming pools or small rivers, typically using the Bambi Fire Bucket.In the picture, a hotspot is getting a good dousing (hotspots, where the fire is still smouldering, can linger underground for weeks).

After the fire, in the (very) early mornings, a FLIR-equipped helicopter detects them and a fireman in the back throws toilet rolls to mark them.

Water sources should ideally be close, upwind and at least the same elevation as the fire, so you don't have to keep climbing with a load on. The Bambi can be transported at quite a speed (up to 80 kts), as long as you don't pitch more than 15° either way.

These days, the minimum machin is likely to be the AStar B2, or a LongRanger with a good engine, with anything lower in performance, such as the 206, being relegated to observation, FLIR or putting out hot spots, since their bucket size is only 90 gallons. In most cases, the killing time is between 30-45 minutes from the start of a fire, so, if you're on standby for Initial Attack, your response time should be as fast as possible (take off inside 3 minutes).

For Initial Attack, you will typically be teamed up with three or four firefighters and their associated gear, which will fill every available hole in the cabin and baggage compartments. In Ontario this is 70 minutes or 120 miles, with 5 crewmembers (a lead and 4 others), with their gear. In Saskatchewan, try the 204 with a 4 person crew.

You will need enough fuel to get them to the spot and be useful while you're there, and get them back again, so, on the way, take note of the nearest airfield or refuelling spot. If the fire gets big enough, they may well bring fuel down in drums, but by that time the bombers could well be there anyway and you will be sent off to another one. Your task, as an initial attack team, is just to stamp on a fire just starting, but you could well be involved in just slowing it

down around people or property. You will get your instructions from the bird dog if it arrives before you get there, and directions will be given with reference to the *head* or *tail* of the fire, which are the downwind and upwind ends, respectively. Left and right flanks are counted from the tail to the head. The tail, that is, the origin, is also called the *heel*).

Here is what you might be looking for on the way there:

What you do when you find your puff of smoke depends on who you are working for. Mostly, you will stay with your team and bucket for as long as you can, until you need fuel (Alberta). Other times, you just dump the team and keep them resupplied until they ask to be pulled out - they will call up water bombers instead of using buckets to dump water (Ontario).

One development of this is *rappelling*, which is the rapid deployment of fire crews by rope from a helicopter (and back in emergency) until the regular crews arrive. This saves them the trek to the fire in the first place and ensures they are not exhausted when they start. They can survive for up to 48 hours in the bush, and a Command Spotter will stay in the helicopter. All this will typically be for fires started by lightning, which are often in remote locations, down to weather conditions roughly equivalent to Special VFR. Rappelling shouldn't be done when it's too windy, or when it's raining, as the special rope will swell up in the pulleys and stop working.

Once a fire has been detected, it will be allocated a number. When it gets beyond a certain size and becomes part of a more serious effort, it will get its own traffic frequency, which should be used by all aircraft entering or leaving, once the bird dog has left for greater things (there will also be different heights to fly in and out).

In fact, reporting for fire duty is a three-step process, starting with the central authority, where you get briefed, allocated a radio, etc., then you are sent down to a lesser centre, where you are finally allocated to a fire. Expect to be given frequencies for it, the name of the fireboss (which will change, as the more expert firefighters get reallocated), the location of the staging area, where you drink coffee when you are not flying, and a map with the coordinates of the landing sites on it.

You will not be the only aircraft about. The combination of lots of smoke (and poor visibility), with heat turbulence and other machines buzzing about could prove to be extremely dangerous - many people report that it's just like being in a war zone (the organised chaos only makes it more so), but if you've ever done the British Grand Prix or joined Biggin Hill circuit you should be alright. Constant communications between machines (on the same frequency) are essential, especially if you are picking up from the same swimming pool (in practice, you will go through the bird dog if you want to change position, if it's still there). One pilot reported that the distance from a pond to the fire was so small as to only require a fast hover taxi between them both, which meant that oil temperatures began to redline, as there wasn't enough airflow to cool things down (sometimes ash will clog the oil cooler). You will be tired, as well, after a couple of days' continuous flying from dawn to dusk, though you won't notice till afterwards, as adrenalin counts for a lot.

Upslope drops should be avoided as much as possible, and only be attempted by experienced crews, especially on low targets, as you will need more airspeed than normal to create a pull-up to clear the area with the load if necessary, without using extra power*. Aside from trying to do a 180-degree pedal turn in a high hover out of ground effect, the resulting high power setting will likely fan the flames, as with a hover drop. It helps if you have a drop off place to one side, and approach with some airspeed, so you can climb with the cyclic, and turn one way or the other with the least power, depending on which way round your blades are going.

*Also watch out for VRS. See *Valley Flying* in Chaoter 5.

With downslope drops, you will not necessarily see the target until you clear the ridge, so you will need targets to line up on beforehand. For very steep slopes, try reducing speed before diving off the ridge, so you don't end up going too fast. Cross-slope drops are OK, if you remember where your rotor disk is. With North American blade rotation, keep downhill slopes on your right, so if the bucket doesn't open or you run out of power, you can drop the collective, put the nose down and be able to use the right pedal to take the strain off the tail rotor. Always approach at a 45°, unless you have a bit of height.

So you don't make embarrassing mistakes, like dropping water in the wrong place, you need to be aware that some fires are deliberately set (see overleaf), to make use of airflow, as in *backfiring*, and there may well be someone around with a driptorch attached to their machine to do it with. *Firing out* is similar, but used more to tidy up a ragged fire edge.

After the excitement is over, and the fire comes under control, typical tasks will be moving men and materials around, putting out hotspots, IR scanning, recovering equipment and generally tidying up.

Bucketing

Picking up water in single-engined helicopters beyond gliding distance from shore has the usual problems, plus possible disorientation if you go too far in. Fast moving streams don't help, making you feel as if you were moving the wrong way, so it's best to find a calm area, as otherwise you will have to move the helicopter to keep up with the water, ending up in a fast taxi unawares - always face the flow of the stream. Approach the water with a slight forward speed and touch the bottom of the bucket to the surface so it tips over and starts to fill as you progress (it otherwise has a tendency to drift around). In addition, your downwash will stay behind you, so you don't get disoriented when it pushes the water away. When the bucket is full, lift it at least mostly out before moving fully forward, so you can check if the wires are twisting (they will untwist more gradually than they would in the air). You will find the wind direction for lifting is critical.

Note: NEVER go over glassy water away from good reference, which does *not* include a radar altimeter or downwash ripples! The safe way is a water landing near the shore, then a hover taxi to the position required (from an experienced overwater pilot).

Be careful if you're longlining with a bucket, as the connector plug for the release is difficult to see and might go under water where it will short out. Longlining would be used where the trees are very tall and there is no water for the hoses nearby - the team might have a small relay tank for you to fill. In fact, this is increasingly popular, as ground-directed water is much more effective than that from the air.

When actually bombing, there is about a second's delay between pressing the button and the liquid reaching its target, which is useful knowledge when you have the bucket at the end of a longline and the darn thing starts swinging. Pulling up (gently!) before releasing the water will help stop this (even on a short line). You need a balance between getting too low and slow over the fire, and risking fanning the flames (and getting ash all over the windscreen, which is wet from picking up the water) and being too high and fast, where the water will just evaporate, although raising the humidity will have a good effect, which is why it's best to err towards the unburnt areas if your target practice is not so hot. High humidity can slow things down enough to get another bucketload in, but I have found that a good dousing is also very effective! Do not hover, except for hotspots.

In an AStar, come in slightly sideways so you can see the fire in the chin bubble. When you see it in the hole in the floor, push the button - the bucket is slightly behind you, so by the time the water  is released, you will be overhead. However, dropping at speed is mostly for cooling after the flames are out, as a fire can stay underground for days - this is when IR is used to detect hotspots. When hover-dropping to cool them down, get the target centrally in the rear inch or so of the floor window (in the centre of the floor shadow on the right of the picture).

A good combination of speed and height is around 60 kts and 50 feet, to be varied according to what is burning. Grass can stand a faster run, since it is quite volatile and sensitive to humidity. Concentration is better in heavier fuels, so you might want to go lower and slower. Don't forget to allow for wind drift!

In valleys, be aware of the extra power required to get you out of the "hole" with a load on - buckets don't always release their load, especially if the connectors get wet when picking up.

Checks for your bucket include the cargo arming switch, bucket open and close switch and electrical and mechanical jettison, and the capacity (for performance). Don't forget the mirror and cables, as for any slung load, and exemptions for going over water without floats, even in a twin, if it's underpowered and can't get back to shore if an engine fails.

Aerial Ignition

The waste product from the activities of slashers, who cut down trees and undergrowth is, not surprisingly, called Slash. The forestry people would normally like this to rot naturally, but many farmers disagree and burn it instead. Because it is both extensive and inaccessible, helicopters are used to set the whole lot on fire, but only in temperatures above freezing, and below 45 kts, from somewhere between 150-300 feet. This technique may also be used when fighting forest fires, for:

- *Temporarily knocking down* large smoke columns so the tankers can see where they're going, by igniting strips of forest within about 10-30 metres of, and parallel to, an active flame front, which starves it of oxygen. Depending on fuel types, temperature, dewpoint, wind velocity, etc. this can last up to around 20 minutes

- *Prevention*. Forests can be torched around priority areas, such as towns or gas plants to get rid of anything that might burn later in the fire season

- *Back burning*, when a rapidly advancing fire is moving towards a cut line, or natural fire break, like a creek, although this could also be done on initial attack. Parallel strips ignited between the fire front and the intended firebreak will drastically diminish a fire's momentum, decreasing its likelihood to spot across the firebreak (where small pieces leap out and start other fires), and keep going, so the firebreak can be used as a defensive line. The idea is to remove fuel from the path of a rapidly moving fire, where the head creates a powerful convective column that pulls air in from all sides, including downwind. When this flow is steady, a fire is started ahead of the main fire from a natural barrier, such as a road or river, which is sucked in and creates a wider firebreak

- *Aiding in ground crew mop up*, by burning off unburned areas, making it safer. Ragged edges can also be straightened, to reduce the perimeter and decrease workload

- *Reducing* the time, effort, and expense of cutting fire perimeters for fires with many fingers, which is very time consuming

- *Steering* a fire somewhere

You might carry a drip torch beneath your machine and get to become an instant pyromaniac with a 20-foot flame about 20-25 feet below the cargo hook, carrying a 55-gallon fuel drum, associated pipes, wiring, etc., and a fire extinguisher to cope with hang fires, or those that start around the valve outlet when it closes (apart from danger to the helicopter, the mixture could fall off and start unwanted fires).

The equipment weighs about 550 lbs when full, and can use straight jet fuel, or have it mixed with a gelling agent, which makes it cling to foliage and be more effective (I love the smell of Petrogel in the morning). There will be a propane flame for ignition. As there is a polarity-switching feature, aircraft rewiring is unnecessary and, although it demands a 25-amp circuit breaker, 30 amps is recommended.

Another way is to use polystyrene balls about an inch across, full of Potassium Permanganate, injected with Ethylene Glycol (anti-freeze) in a special dispenser and ejected from the aircraft. For maximum safety, the balls should not ignite inside ten seconds, but most take about thirty. On larger fires, an intermediate helicopter will be dedicated for the job, on short standby and used for little else, except when prevailing conditions are not conducive to burning, and you may be assigned to other flying duties. Always expect a Forestry Officer to be on board.

Your maximum speed is whatever you can control, after Flight Manual limitations, so that the burning fuel reaches the ground at about 15 kts. If you go too fast, say over 20, the mixture will be too lean and the fuel will be used up before reaching the ground. When greater than translation, fuel will recirculate to give a poor pattern, aside from damaging the equipment. Otherwise, common-sense rules apply, such as keeping the helicopter (or the mixing area) well away from buildings, etc. on the pad, grounding everything, wear flameproof or resistant clothing and equipment, or at least cotton protective clothing, brief anything that moves, etc. Don't forget spares, and to fly into wind, or you will be blinded by your own smoke.

Vertical reference skills are needed, as you must place your torch right next to a second fuel/gel drum for quick and easy turnarounds. In some cases, such as backburning to a seismic line, you will need to be quite accurate, as when igniting the very edge of the upwind side of a given cut line (ensure your mirror(s) are properly adjusted). You must be able to clearly see where the flaming fuel/gel is contacting the ground. Beware of violent vertical up and down drafts close to a large flame front. In an updraft,

keep the disc loaded for the inevitable downdraft that will follow, in which case you should be light enough to allow a safe margin of power to prevent contact of your torch with the treetops. For this reason, 400-450 lbs of fuel seems to work well with a LongRanger. As with any fire, be prepared for moderate turbulence and rapid reductions in visibility, keep your doors on, and your windows closed, as small bits of burning ash and debris can enter the cabin. Be wary of pinstriping, or lettering, etc. that is not painted on (as from the fire authorities).

Activate the torch in spurts from 3-5 seconds, then off for 2 seconds, then on, etc. This prevents pump, and output nozzles, from overheating, and will reduce the chances of the circuit breaker popping. Disarm the circuit breaker during the ferry to or from the burn area, and when the equipment is active, try to fly only over that location, or you might start another fire somewhere else. Carry your Bambi Bucket with you, so if it all goes pear-shaped, you can put the fire out you just started!

EMS/AIR AMBULANCE

Note: Patients just being moved from one hospital to another do not qualify for any exemptions for saving life, but special provisions may apply for duty hours.

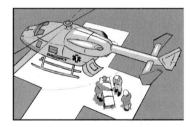

In UK, a *Helicopter Emergency Medical Service* (HEMS) flight is for *immediate* and *rapid transportation* of medical personnel, supplies (equipment, blood, organs, drugs) or ill or injured persons and anyone else directly involved. An approval is required. The purpose of a *casevac* is to give immediate assistance to sick or injured people in life threatening circumstances, typically from the scene of an accident. Both definitions (HEMS and Casevac) mean more or less the same thing in terms of saving life. If the patient is likely to die, you pick them up. The judgment as to whether this will happen is down to the paramedic.

The magic word that differentiates one from the other is *intent*. If you intend to take money off the ambulance trust and/or roster a paramedic to fly with you, you are a HEMS outfit, and legislated for by Part OPS. Casevacs come under the Air Navigation Order.

Otherwise, there are two types of ambulance flight, *Intensive Care Transport* (ICT) and *Ambulance Taxi Transport* (ATT). Both are usually planned in advance, meaning that there's not so much of a rush. Police observers or medical attendants should be able to monitor and assist the patient during the flight and inform you of any problems.

Whatever you get involved in, the following should generally be avoided:

- Previous or present signs or symptoms of epilepsy or other form of fit

- Unconscious patients, unless in-flight attention is available

- Severe haemorrhagic injuries, unless in-flight attention is available

- Abdominal or chest injuries with altitude changes of up to 1500 feet

- Those under the affluence of incohol or drugs, unless prescribed by a doctor

- Persons of unsound mind or who may be a danger to the aircraft or persons therein

"Walking Wounded", that is, passengers who are infirm due to age, ill health or otherwise, may be carried subject to the approval of the qualified medical personnel who should accompany and be responsible for them. No patients should be placed near emergency exits, and wheelchairs, etc. should not impede escape paths. If your Company does a lot of medical work, it may be worth retaining a doctor to advise on certain cases, especially where infectious diseases are concerned. Routes should be planned to take into account changes of altitude and rates of descent, and you will need to accelerate or decelerate with care.

Aside from the patient's condition, the consent of both referring and receiving hospitals is required, together with confirmed arrangements for road transport at the departure and destination airports. You also need to make sure that the type of aircraft is what is wanted, together with the details of staff and equipment.

Specialised equipment should be properly installed, and instructions must be available to all attendants. Some of it could actually be classed as Dangerous Goods (say, large quantities of Aeromedical Oxygen), so you may need an exemption to carry it. Anything that needs to be fixed to the aircraft (e.g. stretchers), or connected to its systems, must be through an approved system, as it must be compatible with the aircraft environment (for instance, that used in road ambulances may be unsuitable for flight).

Oxygen (if used) must cover loading, flight, technical stops and unloading, plus about 1½ hours' reserve.

Patients who can sit should occupy a seat and use a seat belt, which may be a little looser than normally required, except for takeoff, landing, flight in turbulence or during manoeuvres. Stretcher patients should be secured by belts or a harness as per the Flight Manual. Other passengers must be limited to whatever seats are left after stretchers have been fitted.

Flight attendants must have proper experience, and doctors must be qualified and registered in the States concerned in the move (in fact, a doctor is always required on ICT flights and on ATT flights which are other than simple escort cases). Nurses must also be registered in the relevant States and work under direct instructions of an accompanying doctor (or apply those given before flight by one). Nurses must not be used on their own on ICT flights or with any possibility of the patient's condition changing. Paramedics may be employed with the agreement of the patient's doctor. Their training may not be valid in some States.

Tip: Don't let a patient die in the machine! The paperwork will keep you on the ground for days! One flying doctor had to do CPR for four hours in an Australian aeroplane to keep a body fresh enough to declare death after it ended up in the ambulance at the destination (rigor mortis would have been a dead giveaway!)

Note The words "the patient is cold" are used instead of saying "the patient is dead", just in case any relatives can lip read, aside from the fact that only a doctor can certify it. The fact that their head might be in a different bag to the body doesn't come into it. They're alive until a doctor says they're not! So if you're told someone is cold, make sure they are not just wanting another blanket, because you will get a surprise if they sit up and ask for one!

PLEASURE FLYING

The idea behind this is to carry groups of people, generally members of the public who have never flown before, for a short period, which can be anything from three to fifteen minutes, depending on whether you're just going round a local show, the Niagara Falls (9 minutes), the local city (30 minutes), or mountains. Here, we're more concerned with the first, which takes place on a more informal basis. Also, the dimensions for landing sites given here are largely applicable to UK, but are still good to use elsewhere.

It can be very lucrative if the operation is slick and smoothly controlled, and it's also an ideal opportunity to promote aviation in general, so everyone should take care to ensure that the customer's association with aircraft (and the Company) is a happy one. The machine must therefore be handled smoothly with no sharp manoeuvres, unless they're specifically asked for, with all passengers being in agreement. The only types of flying generally recommended are spot turns and sideways movement in the hover and normal climb, descent and cruise. Oddly enough, hovering manoeuvres are quite popular, but not good commercially, as you need to give them their money's worth, so you should keep the rhythm going.

Because ground idle isn't officially counted, your turnaround time does not affect the revenue per flying hour, but, obviously, the quicker you are, the more cash you have in your pocket at the end of the day. The duration of the trip itself is what determines productivity - the shorter it is, the more you get, but regard 2½-3 minutes as the minimum before you get complaints.

On top of the public actually flying, there will also be those attending the associated event, so don't forget the exemption to fly close to assemblies of lots of people. Sites must be organised and staffed to afford the maximum safety for all concerned.

The size and location of the site, the type of event and anticipated numbers need to be noted and a site inspection arranged before you start. This must be done by someone who knows the requirements, such as a qualified pilot, because positioning must be paid for as well, and if the site turns out to be unsuitable and is rejected on the day, you've all wasted your time.

Look at a site's physical characteristics and position relative to areas available for emergency vehicles, together with those for the aircraft. For instance, it may be acceptable for crosswind takeoffs (if performance is OK) if emergency vehicle access is better in that direction. You are also allowed some discretion if an otherwise perfect location is spoiled by one or two major obstacles which can simply be avoided by curving the flight path (there might be a tree in the middle), but an airfield should be alright on this point. The authorities should be notified at least 7 days beforehand, as must emergency services.

Running The Site

On the day, you should check your area is roped off properly, and is the same as was agreed originally - beware of tents and marquees creeping up on you. If you don't use the agreed area, then (by arrangement) you will have to inform the authorities within a few days as they will have been informed of the original plans.

Next, find the organisers and check for other activities that may affect you, like aerobatics, balloons or parachuting (rotors must be stopped for the latter in UK). Give them an idea of what you want the announcements to sound like (preferably every half hour) about your activities, and let them know your start time. If they have free seats, you will need a positive means of identifying whoever gets them. You may be plagued by people claiming they're from the organisers or the local papers, but unless you can identify them, politely refer them back to the organiser.

Meanwhile, back at the Operational Area, set out the safety equipment just inside the ropes and carry out whatever checks you need on the fuelling gear - this will save time later when the pressure's on. Show the marshallers around everything (it's a good idea to keep the emergency equipment accessible, but out of sight, as the public tends to be put off by the sight of anything designed to help in emergencies, like checklists). If local fire engines are on site, rescue equipment is not necessary but take it anyway, because they will either be swamped with children sounding the sirens or be called away and you will have to stop flying until they come back.

Standard Rescue Equipment

For the UK, at least, this consists of something like the following:

- A vehicle capable of carrying everything - not a wheeled trolley, but something self-propelled. A car or van will do, but it must be able to go over all relevant surfaces. The ultimate is a long-wheelbase LandRover, as you will discover when you try and pull a trailer with all the stuff mentioned below with it, not to mention fuel barrels.

- 11 kg Dry Chemical fire extinguishing agent, 1 x 7.5 kg CO_2 or 1 x 3.5 kg BCF extinguisher and 1 x 20 gal premixed AFFF foam unit with a minimum discharge rate of 16 gallons per minute. Although the BCF extinguisher is as good as the first two combined, the chemical is difficult to get. Extinguishers have to be serviced every year and tested every 10, according to the BSI.

- For each marshaller, helmets with visors, flame resistant gloves, fire tunics or donkey-type jackets and stout boots. Most local fire brigades have a surplus equipment office where you can buy them.

- Release tools as follows:
  - 1 axe (rescue, small, non-wedging or aircraft)
  - 1 x 24 inch bolt cropper
  - 1 x 40 inch crowbar
  - 1 harness knife
  - 1 flame resistant blanket

- Medical equipment as follows:
  - 6 BPC9 dressings or equivalent
  - 6 BPC12 dressings or equivalent
  - 6 triangular bandages
  - 6 foil blankets
  - 1 pair scissors
  - 1 basic First Aid kit
  - 2 stretchers

BPC9 and 12 are now officially out of date, but still about, so try to get the right ones, as insurance companies will do their best to weasel out of any claims, and you don't want to give them an excuse. Scissors are in the average First Aid kit, anyway.

There should be only one entry and exit to the operational area, usually under the control of the cashier, but there's a danger of the money being taken if an emergency crops up and it's left unattended, so you need a hefty table with a very large metal box screwed down to it, padlocked, with a slot in the top into which the money goes.

The minimum people needed to run the site effectively will be 3, one to collect money and brief passengers (the cashier) and the remaining 2 to marshal passengers in and out and operate seat belts, etc. If, for any reason, such as last minute sickness, you can't get enough people, you can get away with one marshaller on passenger movement, if

all embarkation and disembarkation is done from one side of the aircraft, one door at a time. It's not recommended, however, as it takes longer and passengers en masse must be regarded as thick as two planks - they will take every opportunity to walk into a tail rotor, regardless of how many warnings you give them. A large version of the briefing card is recommended as something they can ignore in the queue, on top of the ones you hand out anyway. It'll be a waste of time, but make the effort.

While marshallers can also be the rescue crew, they're not expected to wear firemen's uniforms all the time, but should still be dressed well and in a good enough substitute if something happens quickly (so no shorts and T-shirts). A good source is Air Cadets, who not only look smart in their uniforms, but are also keen to be near aircraft, and will do a day's work for free flights, which is where you can use up any positioning flights, as you can't sell them to the public. That's not to say that you should abuse the privilege, though. Make sure they get their money's worth.

Place the sign with the Company identification (and the price) on about ten yards or so from the cashier's desk, so if people are put off by the charge, you don't give the wrong impression by having lots of them turning away at the last minute. It also saves the cashier answering the same questions all day. You will need plenty of other signs around the event as, unless you're careful, regulations will ensure that you're far enough away for people to think you're nothing to do with it. *Potential passengers will not walk more than about 100 metres.* As with any customer, the sales process must be made as brainless as possible, as pleasure flying is done on impulse 99% of the time. If you make it difficult for them, they will not do it, so make sure you can take all major credit cards, so they don't feel like they're spending real money. Next you should brief the loaders, ensuring that they know that people should always approach and leave the helicopter from the front and that nobody should be allowed further aft than the rear skid support. They also need to know about the opening and closing of doors and the operation of seat belts, all of which should be covered anyway in the passenger brief.

The most dangerous time is when the passengers change over, so that's when marshallers must be most wary. When you land, outgoing passengers should be out of the area (or at least the edge of the rotor disk) before the others are ushered in, although you can tighten things by shepherding the new ones to the edge of the disk while the others are getting out. Never close the throttle to ground idle during this, so you can lift into the air and get out of the way of anyone you see about to run round the back end - believe me, they will! The tail rotor is dangerous at whatever speed it's going.

With reference to refuelling, it's very tempting to carry on till the last minute with a long queue, but be careful about your fuel reserves. Not only is it good airmanship to land with a reasonable amount on board (don't trust those gauges!), but you must have a 30 minute break every 3 hours anyway (in Europe). Passengers (and employers) understand a helicopter stopping for fuel, but not for you sliding off for a hamburger somewhere.

There's a safety point as well. In a way, helicopters are regarded in the same sense as a fire engine - the public make no distinction between an old one on show or a new one actually on duty. If there's a fire, they will turn to anything for help. The same goes for a helicopter. If an accident happens, you could be asked to ferry someone to hospital. Do you know where the nearest one with a helipad is, and will you have enough fuel to get there?

For maximum revenue, fill the machine up on every lift. If it takes 4 passengers, don't fly with less than 3. If only 2 turn up, they should wait, or come back later when others have arrived. Children less than 2 years old on an adult's knee should have an approved seat belt extender provided for them - don't expect to carry more than one, or preferably none at all, because they're too young to appreciate it and you stand a real chance of putting them off flying for life, aside from being sick down the back of your neck. Don't sell more than 2 loads in advance in case something happens and you have to return all the money. Also, don't strap one passenger in whilst waiting for more custom. This is for two reasons; firstly if nobody else comes along you're obliged to go up with just one person (uneconomical) and, secondly, you will have to make conversation by shouting while the customer is waiting, because usually you're the only one with a headset (talking of economical flying, and the subject of freebies, if the show organisers send along more than one, allow them up one at a time, so the costs are covered by the other revenue passengers on each trip; if you take all the freebies up at once, you lose money on the whole lift).

You'll need to identify those who have paid, which is usually done by sticking labels on them or date-stamping their hands. If someone asks how long the flight is (they will), say six miles or so - it sounds better than three minutes. While the aircraft is flying, ground staff can brief the next load like this:

"When the helicopter lands, please stay here until you are called forward, as we have to unload the other passengers first. You and you

go to the right hand side as you look at it, one to the front and one to the back door. The other two please go to the left hand side, you to the front and you to the back door.

Note: Don't mention "the back" by itself or they will take it literally.

When you get in, please do not step on the floats, but use the foot rests on the skids which will be pointed out to you. Once you are in, we will do up the seat belts and close the doors. After you land, we'll get you out, so just sit tight and wait for us.

Some very important safety points - please don't touch the door handles in flight, don't throw anything out of the windows, and keep away from the tail rotor - always move towards the front where the pilot can see you."

Again, this sort of stuff should be on the standard Passenger Briefing Leaflet - you could hand out a few to keep people in the queue occupied, as they won't listen to you properly, anyway. Have one enlarged and pinned to a large board so it can be read from a distance. The cashier will need some change, but you could keep the price at a round figure so you don't need it anyway (try multiples of 5). When it's closing time and obvious that not everyone will get a trip, stop selling in good time.

Keep an eye out for your Inspector! He will be lurking behind a tree with a notebook.

Site Dimensions (for UK)

The Operational Area, which is under positive control of the Company, encompasses the Landing Site, the taxiways, HAAs and IAAs (see below) and takeoff, climb and approach slopes. It has side surfaces rising upwards and outwards to 100 feet at a gradient of 1:1 from its edges, unpenetrated by obstacles and will be fenced, roped off or otherwise protected from intrusion by unauthorised persons, so it should not include a right-of-way. Rope and stakes for demarcation and public control are not obstacles for this purpose.

Note: The primary concern is that the helicopter must be able, at all times, to carry out a safe forced landing if an engine fails. This requires an obstacle-free corridor to be available the start of takeoff to the end of the landing.

FINAL APPROACH AND TAKEOFF AREA (FATO)

This is inside the Safety Area, where the final phase of the approach to hover or landing is completed, and from which takeoff is started. It may be square or circular and minimum width is 1½ x D or 24 metres, whichever is the greater. Mean slope in any direction should be 3%.

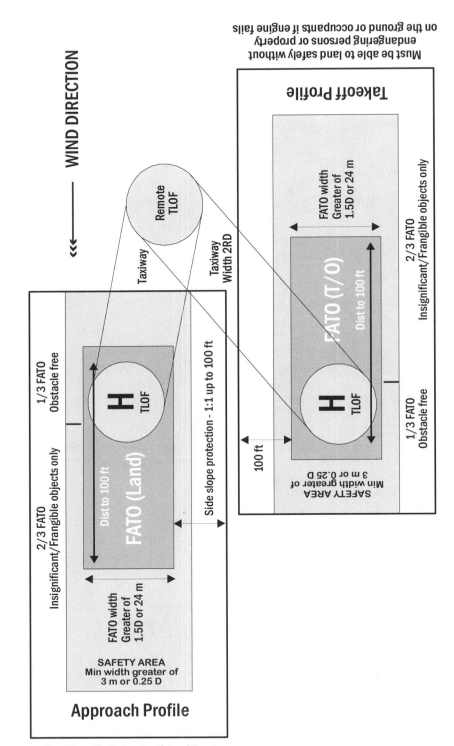

WIND DIRECTION

Must be able to land safely without
endangering persons or property
on the ground or occupants if engine fails

Takeoff Profile

FATO width
Greater of
1.5D or 24 m

FATO (T/O)
Dist to 100 ft

2/3 FATO
Insignificant/Frangible objects only

1/3 FATO
Obstacle free

100 ft

SAFETY AREA
Min width greater of
3 m or 0.25 D

Remote
TLOF

Taxiway

Taxiway
Width 2RD

H
TLOF

H
TLOF

Side slope protection - 1:1 up to 100 ft

1/3 FATO
Obstacle free

2/3 FATO
Insignificant/Frangible objects only

FATO (Land)
Dist to 100 ft

FATO width
Greater of
1.5D or 24 m

SAFETY AREA
Min width greater of
3 m or 0.25 D

Approach Profile

Must be able to land safely without
endangering persons or property
on the ground or occupants if engine fails

SAFETY AREA (SA)

Surrounds the FATO, for reducing damage to helicopters accidentally diverging from it. It is also square or circular, according to the FATO, and the minimum size is 3m or 0.25D, whichever is the greater. It must be free from obstacles. The combined size of the FATO and SA should be at least twice the overall length, including rotors, of the helicopter and, where it contains the TLOF (see below), the surface should be firm and not blow away with downwash. The mean slope may be 4% from the edge of the FATO.

TOUCHDOWN AND LIFTOFF AREA (TLOF)

A load bearing area for touchdown and lift off, free from slopes, to ease passenger embarkation and disembarkation, and not have them walking uphill. The TLOF has a diameter of at least 0.83D and may be contained within the FATO (it's the same minimum size), but it may also be separate. The TLOF should be flat, but maximum slope is 2%.

TAXIWAYS

Used where the TLOF is remote from the FATO, with a minimum width of 2 rotor diameters. The maximum slope should not exceed 10% transversely and 7% longitudinally, within Flight Manual limits.

SIDE SURFACES AND SLOPES

The Safety Area has side surfaces rising upwards and outwards to 100 feet at a gradient of 1:1 from the edge, not penetrated by any obstacle.

OVER WATER

Flight over large areas of water shall normally be avoided but, should it occur, the aircraft shall be fitted with floats & jettisonable emergency exits. Lifejackets shall be worn and a radio equipped safety boat will be available, all of which should form part of the briefing of the passengers.

Note: Performance figures here are for a Bell 206B at +20°C and 1000 ft PA at max AUW (hot Bank Holidays).

Note: The TODAH and IAA are essentially the same patch of ground, except that the TODAH is slightly longer. As both the HAA and the upwind third of the IAA must be obstacle-free and are at opposite ends of this area, the whole TODAH should be obstacle-free.

SPECIAL EVENTS (FOR UK)

Vast amounts of people being moved into a major sporting event (such as The british Grand Prix) make the feeder sites used for their lifting and dropping off liable for special treatment. These events are good for business - one good day at Silverstone keeps some companies in profit for the year. As for Pleasure Flying sites, the authorities need to be notified (in this case at least 28 days before), but other considerations arise as well:

- First of all, if you sell single seats to the public, rather than the whole capacity of the aircraft on a "sole use" charter basis, you will either need a full Air Transport Licence, or an exemption (pleasure flying is a special case). It also needs to be done in your own right; you can't do it on the back of someone else, as you can with an AOC sometimes. Again, there is a special form to fill in which will cut out most of the lack of communication over this subject, and you should find a copy in Ops.

- Secondly, you will need to arrange arrival and departure slots, which are usually at a premium. Because of the numbers of aircraft involved (usually over 126 H1 types alone at the Grand Prix), there will be a briefing for all concerned well before the event, at which all companies are expected to send a representative. At the very least a NOTAM will be issued (H1 helicopters, by the way, are less than 15m long, and H2s between 15-24m; they therefore require different treatment at their feeder sites).

A feeder site is one where more than five movements take place in any one day in connection with an event, as a result of which they require special facilities (a movement is a takeoff or a landing). If using H1s you can get away with normal equipment as used for pleasure flying, but H2s need something a bit more macho. Actually, it's basically the same, but the vehicle must have four-wheel drive and there must be a minimum of 60 gallons of water and 5 gallons of foam concentrate, with equipment able to deliver it at 40 gallons per minute. A minimum of 100 lbs of CO_2 or 50 lbs of dry powder or BCF is also required.

The rescue and medical equipment requirements are also more comprehensive, needing transfusion and resuscitator gear to be readily available as well as the usual suspects.

:

LINE PATROL

It's pretty hard to get lost doing this! Helicopters are used to check power lines because it is way cheaper and faster than getting linemen to do it on foot patrol (and it doesn't damage the landscape). In addition, certain types of damage can only be seen from above, and tree trimming assessments are more accurate from the air. In fact, aside from normal movement of people and material, there are four ways in which helicopters are useful here:

- *Routine Patrol*, where 7 main fault groups are looked for and reported upon.

- *Thermal Patrol*, where IR is used to detect faults.

- *Slinging*, either lifting and delivering poles or stringing cable.

- *Emergency and Post Fault* - looking for trouble after it has happened - emergencies due to weather may also make road access impossible.

Unless you are a specialist company, you will be carrying observers from the Electricity Boards, who are fully and professionally trained to exacting standards - they need to be, as following and inspecting tower lines calls for a high degree of proficiency and concentration from everyone. The very nature of the exercise (flying close to the lines inspected) means that, for most of the time, you will be very near the avoid curve. In fact, you will be flying at such a speed that, if the engine fails, you will be going nowhere but down, which is quite interesting when the line runs through trees and you have to decide whether you're better off in them or dodging the cables (refer to the *Engine Failure & Autorotation* discussion in *Techie Stuff*. An engineer once told me that, if you went into the wires, at least they would know where you were as the trips for that location would go off).

As you fly along, the observer(s) will take notes, or speak into a machine, while looking for problems such as bird or animal damage, broken stays or insulators, lightning damage, trees or other risk elements encroaching on the lines, all in the space of a few seconds' pass, which is why the smaller poles are more difficult, as they come up so quickly. On the left is a typical example of tree overgrowth (the lines are just below the middle, on the left). Once the flight is finished, the tape is transcribed into language that linemen understand, filled in on a form and shipped off to the relevant departments for further action. However, if something serious is seen, an immediate report can still be made to the local control room.

For normal wooden poles, being one and a half rotor spans laterally and flying at about 25 kts is the ideal, and probably the limit of the average exemption. At the very least, you need a positive airspeed, that is, one showing on the ASI (if your company does not regularly check power lines, the exemptions may specify at least 45 mph). The authorities assume the flight won't take place in the avoid area, so exemptions, etc. are geared towards looking after third parties on the ground, but the flights are still Commercial Air Transport (in UK), and entering the avoid curve is therefore prohibited, which is where a twin capable of Cat A performance comes in useful (under EASA, however, line patrol is aerial work). Hovering should be avoided in any case.

Note: The *absolute minimum* for this job is a Bell 206!

No flights should be made at night, over fuel installations or congested areas of any city, town or settlement. Dispensations will be required to fly near nuclear installations and prisons, just in case they think you're part of an escape plan. Flights should also be confined to within 300 feet of the lines concerned (but no closer than 1½ rotor diameters to the lines when level, or 1 diameter plus 30 feet when above). Flying above and to the right is most preferable, so the observers can see, but sometimes you have to flip over to the other side for safety, or avoiding animals or houses. Lines should also (normally) be crossed vertically at least 100 feet above them, over a pylon. Even a specialist company will not be allowed closer than 100 feet to any people or vehicles directly concerned with line operations and 200 feet to any other structures than those to do with the lines themselves. You shouldn't, but if you must go under a wire, get your skids on the ground as near to a pylon as possible, where the ground clearance is greatest. A reference ahead is useful, as is one to the side. Looking up as you go under is not a good idea, as it could cause an unconscious climb.

If this page is a photocopy, it is not authorised!

As well as the proper permissions, other problems include insurance. Whatever you get should also cover frightened animals bolting and causing havoc - this usually happens with sharply changing noise levels caused by rapid manoeuvres. If you can't help flying over animals, at least try not to chase them through the fence! With horses, go up, without banking or causing blade slap. Once you are at 200 ft, they shouldn't move. Cows generally don't bother, but when they do, prefer you to be as low as possible, because they can't get their head high enough to see what's going on. If you have to go around animals, make sure they run towards open ground.

The University of Bristol carried out trials in 1963 and 1964 on their own farm, using a Bell 47 against horses, cows-in-calf, heifers, in-lamb ewes, and chickens and cows inside buildings. Experienced stockmen were used, and observations were also made on the production of milk or eggs. Passes were made by the helicopter at 60 and 35 feet, at speeds up to 25 mph. It was concluded that completely housed farm stock is not affected at all (although you would still be advised to avoid such buildings wherever possible). When out of doors, reactions are very temporary, after a fleeting period of bewilderment when the animals could injure themselves by hitting fences or falling into ditches. Poultry (and ostriches!) out of doors, however, present the most problems, and will react even to the helicopter's shadow.

Line patrol should not normally happen if the visibility is less than about 1 mile, and 2 if raining (1 mile is around 6 towers or poles ahead). These limits are higher than usual because moisture sticks to the windscreen at slower speeds, and things are further complicated if you have no windscreen wipers, as precipitation won't blow away either. Under those conditions, speeding up to get rid of water is not what you want to be doing, especially when those grey towers merge into the weather (in fact, hovering gets rid of water best). Give serious consideration to aborting if there is a lot of rain. Also, line patrol should not normally be attempted if the wind is above 25 knots, and the cloudbase below 400 feet. As it happens, the observers will probably call things off before you do, as wires tend to merge into the background when it gets a bit grey and horrible, and they won't be able to do their job.

Always try to follow the line as near into wind as possible, or, if not, in trim at least, which will help if you lose tail rotor authority. If it's around 10 knots or so, being downwind generally will only ensure the transit time along the wire is too fast, with the consequent danger of you trying to slow down and having no airspeed - if more it may be rough as well, especially in the mountains. Monitor the instruments and be particularly aware of overtorquing or overtemping if you are trying to stop yourself being blown towards somewhere you don't want to be, like a set of tower lines. Don't forget tail rotor and wire strikes, and other lines (especially tower lines) crossing - the observers may be too busy to assist your lookout. If a closer inspection is called for, DO NOT come to the hover and backtrack, but gain height and speed, positively identify the area and make a conventional circuit and approach to come to the hover alongside the line into wind.

A constant lookout must be maintained at all times, especially for fast, low flying military aircraft, but if they are as low as you are, they are the ones in trouble! *High Intensity Strobe Lights* (HISLs), nav lights, landing lights and anti-collision beacons must be on at all times. HISLs should be at least 2000 candlepower (so don't drop one or you'll have to pick them all up). If they become unserviceable, patrol above 500 feet. Don't plan on doing more than 5 hours per day due to the high workload.

Wires

70% of wirestrikes happen with highly experienced pilots, and around 50% happen in clear sky conditions. Very often, a wire that is seen on the way into a landing spot is either missed or forgotten about on the way out - it is therefore a good practice to circle a spot on the way in, to give yourself the best chance of catching any wires that may be present, as they are usually below the resolving power of the eye. If you cannot see wires themselves (they are below the resolving power of the eye), there are often clues to their existence. When following a power line, for example, if the line changes direction by more than a few degrees, there may be guy wires to help hold the pole erect where it changes. If there is a road in the way, the guy wire may well go across it, according to convenience. Not that you should be that close to have it affect you. If a wire strike is imminent, try not to flare - you will simply be presenting more of the airframe to the wire, but it's very difficult not to touch the cyclic!

Power Line Cleaning & Maintenance

This is done when the electricity people can't get to them by road. As you can imagine, a lot of training is required, especially when the lines are kept live. Your hovering needs to be precise for long periods. For cleaning or deicing, a water pressure system powered by a small gasoline engine is used, and you get to hover a bit further away. There is a platform on your helicopter, on which the lineman sits, who must first connect the machine to the line to equalise the potential. This means the helicopter is live as well, so special maintenance is carried out beforehand to ensure the ship's normal electrics aren't fried. However, line construction involves (usually) 2 lineman standing on a skid, to change the clip holding the line to a pulley so it can be pulled through and changed to whatever line they need. Watch the markers when close to the tower - if they are moving, your rotor is too close and could hit if the lineman pulls hard on the tower. Keep your tail clear!

PIPELINE SURVEY

This is similar to Line Patrol, so most of what is said there will apply here. Gas pipelines (or whatever) are not very far below the surface of the ground, and potential hazards include building works, ditch construction, drainage, flooding, leakage and falling trees. 1/250,000 scale maps should be provided, on which the pipeline route should be marked, though it will be obvious when it's just been laid. Observers will be using larger scale maps supplied by their companies. The normal patrol height is 300 feet (in UK, anyway), with an associated speed of 50-80 kts, taking into account the efficiency of inspection, terrain, wind direction and practical helicopter operation (avoid curves, etc.), but between 500-700 is recommended to avoid low flying military aircraft. You should not approach closer than 300 feet to any person, vessel, or structure, and only closer than 500 feet within a horizontal distance of 300 feet from the pipelines. The cloudbase should be at least 200 feet above inspection height, with a minimum visibility of 1500 feet.

UK POLICE OPERATIONS

A police force will either own its own aircraft or charter from operators as and when required, in which case the only things you can get away with are low flying (closer than 500 feet, etc.) and going in bad weather. Even then, you can't get closer than 50 feet, or 1½ rotor diameters, whichever is greater. In fact, to take full advantage of any restrictions, your passenger needs to be either a police officer, an employee of the police authority, a medical attendant, a pilot under training, an Inspector, a Fire Officer, a Customs Officer on a joint operation, or any other agreed in writing. However, there will always be a police observer, and the job involves a lot of cross-controlling, not to mention short-notice tasking. As they may need to recover their costs from time to time, they may also need a Police AOC and Ops Manual (PAOM), though this depends on the country.

The accepted rule in Police Air Support is that the aircraft (with all its high tech equipment) can only be truly effective if it can be overhead the scene of an incident inside 15-20 minutes. After that the chances of the criminal still being around to catch are minimal.

Prisoners

A prisoner is a passenger and qualifies for the normal safety considerations, although their movement is definitely not routine. Handcuffing should always be done to the front, so seat belts can be released in emergencies. Potentially violent prisoners should be carried one at a time and have enough escorts to restrain them. In any case, at least two escorts should be used, ensuring that neither the pilot, the controls or the exits can be reached (no prisoner should ever be behind the pilot). If a prisoner does become violent, land as soon as you can and have them continue by surface means.

Persons under the Influence

These should be avoided, but sometimes it can't be helped. You need to decide whether they are fit for a normal seat or need a stretcher. You will also need at least two escorts for restraint and emergency evacuation, and a suitable receptacle for vomit. As with prisoners, land as soon as practicable if there are signs of violence. Further movement should be undertaken by surface means.

Police Dogs

Should be embarked or disembarked with the aircraft shut down, but if this is not possible, the observer should meet the handler and dog clear of the aircraft for the briefing (make sure the dog reads the card). The dog should be on a short lead, so it doesn't interfere with anything. Fly smoothly, but be prepared to land if the dog becomes unwell, although you will find that they mostly like travelling and prefer to look out of the window rather than being made to lie down. If the dog breaks free and goes for the tail rotor, the handler must not attempt to follow, but give you a signal to close down. Don't let dogs in the cabin when they are wet! Or rather, let them shake themselves dry outside, otherwise the windscreen will get wet all over the inside.

If this page is a photocopy, it is not authorised!

Weapons and Munitions

The overriding consideration is eliminating danger to the aircraft, its occupants and persons and property on the ground. Munitions include gas/smoke canisters, stun grenades, shotgun cartridges and ammunition for rifles and sidearms. These should not be in a loaded state:

- double- or single-barrelled shotgun (unless automatic or pump action)

- baton or dart gun

- CS discharger

Loading, unloading or firing of weapons on aircraft is not allowed. Where all passengers are securely seated throughout the flight, loaded weapons must be in a safe condition, with weapons and munitions in holds, compartments, or other areas inaccessible in flight, and secured normally. Weapons or munitions must not be distributed until the aircraft has landed.

Weapon	Safe Condition
Self-loading pistol, self-loading rifle, carbine, automatic shotgun or pump action shotgun, bolt action rifle, automatic rifle.	Working parts forward and trigger released, safety catch applied where possible, magazine charged with ammunition and fitted to the weapon. NO ROUND IN THE BREACH.
Revolver	Cylinder loaded with ammunition; weapon in a secure holster, to prevent accidental discharge.

When passengers on a special operation need their weapons handy, hand weapons and spare ammunition for them may be carried in readily accessible boxes or holsters, with the ammunition in pockets. Rifles and shotguns may be stowed securely within the cabin, with spare ammunition in body belts or readily accessible boxes. Gas/smoke canisters must be in boxes but these may be readily accessible. Boxes must be strongly constructed, fire resistant and have an "explosives" label.

Bodies and Remains

Their carriage is affected by how inaccessible they are, that is, precluding other methods of transport. The main considerations are the health and hygiene of the aircraft occupants, which means they might have to be carried outside if they are a bit ripe, so you need to be current on winching or slung loads. Bodies and remains should be in body bags or coffins. Any spillage of body fluids needs a thorough wash down as soon as possible.

Formation Flying

Used when chasing another aircraft, but not at night, in cloud, or when the cloud base is below 500 ft or visibility below 3 km. As the pilot of the other machine may not be aware of your intentions, and might not even wish to be identified, you shouldn't do this too closely. In other words, spend the shortest possible time at the minimum permitted distance from the other machine, and shadow from the maximum range consistent with getting photographs, or other evidence, and maintaining visual contact - one official definition of formation flying is "flight by sole reference to another aircraft".

Do not endanger the other machine or attempt to force it to alter height or heading, or to land, because you may need to take avoiding action if the other machine endangers you. If you cannot establish RT contact, approach the target in straight flight, which may be level, or climbing, or descending. Establish a stand-off position behind, between 4 and 8 o'clock, not closer than 200m. Reduce the range slowly and progressively to at least 100m, moving back out once the required evidence of registration, type and other features have been noted.

In the US and Canada, prior planning and briefing between crews are required before formation flying occurs, and it is very strongly recommended that formation flying between different kinds of aircraft (helicopter and aeroplane) does not happen. Several accidents have proven that it is not a good idea.

Landing Helicopters on Roads

The area must first be secured (but see below), with radio or verbal communication having taken place to confirm that you have authority to land there. Unaffected carriageways should be closed. Normally, you can only land at an unsecured site (i.e. where the police are not in attendance) in remote rural areas outside congested areas, but you may do so on dual carriageways (divided highways) or motorways (turnpikes) in daylight, where no traffic is moving - no landings should be made by night. In any case, there must be no threat to persons or vehicles on the ground from the helicopter, or vice versa. Always be aware of the effects of your downwash, which may blow away crucial evidence, not to mention dislodging broken glass and other loose debris, particularly from damaged buildings, at bomb scenes.

Try not to land in school or other play grounds, or areas where children might be confined or suddenly emerge. Don't use the aircraft presence or public address system to clear children from a site.

HOVER EMPLANING AND DEPLANING

This is defined as allowing trained persons to enter or leave a helicopter without its full weight on the ground, and done where you can't land properly. There should be no danger to third parties and minimal risk to the aircraft, crew, seated passengers and those carrying out the activity. (the major consideration is engine failure). Usually, being one or two inches off the ground is enough, but sometimes you might have to go up a couple of feet. Doing it from 8' (as favoured by Forestry) is illegal in Canada, because CARs requires a *low* hover where people can step out - normal is 4' for a 206 and 5' for an AStar.

It should be done in day VFR, with one skid or wheel in contact with the ground to get rid of static. One passenger should be seated before the next gets on board. To give you a decent power margin, the weight should not greater than 95% of the maximum, and you should have at least a 15% power reserve in the low hover anyway. The time in the hover should not be more than the time limit for take-off power. Cargo should not shift.

Your safety briefings should include the effects of C of G changes, especially when it comes to toe-ins. Make sure your passengers transfer their weight slowly (no jumping off, and one at a time!) and not move the helicopter when shifting their baggage or closing doors. Here's a typical example from a senior pilot:

> *"Basically I have to hold a 500, left skid on top of a 6' stump while a 240lb cutter climbs out the back seat with his saw, sawgas and lunchpack etc. in his hands. Then put the other skid on another stump while the front and rear seat dudes both scramble out together with all their gear - same basic deal with the hookers. Takes about 30 minutes all up, and by now I'm a nervous wreck and we haven't started work yet!"*

A crew leader should be appointed, who will co-ordinate the procedure with the passengers and crew, normally being the last one out. If a gust of wind causes the aircraft to move, passengers should stay in if they are mostly in, or get out if they are mostly out.

DEPLANING PROCEDURE

Seat belts must be unbuckled as instructed, then rebuckled. Only one person or item of cargo should leave the helicopter at a time, from the left side (Bell 206), with the weights being gently transferred to the ground. One person should ensure that all doors are closed after everything has been unloaded. People and cargo should remain huddled ahead of the passenger door where you can see them until you depart.

EMPLANING PROCEDURE

The crews doing this should be trained to your company's specifications. Some companies require 3 hover exits per crew *and* pilot (e.g. HTSC), whilst others only require one (CHL) (see the Company Ops Manual).

Seat positions should be determined beforehand, with people and cargo huddled in a safe place beside the intended arrival area. People must only approach when signalled thumbs up, or an exaggerated head nod), from the left side, with only one person or item entering the helicopter at a time. Weights should be gently transferred from the ground to the helicopter. Seat belts should be immediately fastened once seated, and doors closed.

PARACHUTE DROPPING

In UK, no parachute dropping should be undertaken unless (as a pilot) you've been approved by the British Parachute Association and the parachutists themselves are in possession of an Operations Manual authorised by them. It's probably similar elsewhere - in Canada, for example, the company must have an Operations Specification that allows parachute jumps, and most companies don't have one.

You get your certificate by passing a check ride with a TRE, who in turn has been approved by the BPA. The normal regulations for the dropping of articles from aircraft also apply. In addition, the Flight Manual should include a Supplement to cover the situation. For some strange reason, parachutists do not seem to be classed as passengers or freight, so it's a good question as to whether a parachute trip is actually Commercial Air Transport or not. Check your insurance, even though on the way down they are not in the aircraft.

Parachutists should be strapped in at all times except immediately prior to dropping, and before takeoff they should be shown how to secure seatbelts so they don't flap around in flight, as part of a proper briefing. There should be no loose articles in the cabin, and seats must be removed, as must dual controls if one intends to drop from the front seat (of a helicopter). There should be no other passengers.

Don't use static lines and remove the doors (check the Flight Manual).

A typical freefall drop needs one pass over the drop site into wind at around 2000 feet, where the jump master will drop weighted paper markers. You then start climbing to the drop height, turning downwind and keeping the markers in view. When at the drop height, come over the site again at about 60 knots into wind, where the jump master will guide you to where he wants to be.

When dropping, use both sides of the aircraft if possible (difficult in a Bell 47), so the lateral centre of gravity limits are not exceeded - this is one of those times when you might want to calculate it for takeoff and landing. The helicopter should be level, above 2000' agl with an airspeed between 20-70 kts IAS.

AERIAL FILMING AND PHOTOGRAPHY

Flights should be planned so that emergencies don't put structures or persons in the vicinity at risk - you must observe the low flying rules unless an exemption has been granted; in practice, you can get down to 200 feet for photography, but you will need to keep a record of when it was taken advantage of, as you would with any other.

If a door needs to be removed, loose articles and surplus seat belts should naturally be secured (or removed - check the Flight Manual) and manoeuvres carried out where possible so the side of the aircraft without the door is uppermost; people near the open door should wear a bit more than the seat belt supplied. This point is controversial - very often a photographer will expect the door to be off, not have a mount and just use a normal seat belt. No way, José! In this case, I would insist on at least a rope around their middle loosely attached to an anchor point as well, but a professional outfit - which includes your Company, of course! - will have its own despatcher's harness. Think about it - the photographer needs to shift to get a better position, so he undoes the seat belt to help him get it! Then falls out!

A camera mount will normally be fitted by the company supplying it, but you should oversee the work and annotate the loadsheet accordingly (it should not be fitted unless there is a Supplement to the Flight Manual covering its installation). When the mount is in place, the C of A changes to Aerial Work, therefore no passengers should be carried without an exemption, or unless they are essential to the operation, which includes the photographer. The C of G requirements will change as well. If you get a choice, sit the cameraman on your side -

keeping the target inside an area in the top front part of your side window will give him the field of view required.

It's important to get what's wanted first time, not only for economy, but also noise nuisance. Camera crews are famous for wanting "just one more shot" and going "a little lower", but don't push yourself or the machine.

Bear in mind the height/velocity curve - the top for the 206 is 400 ft, but I have taught successful engine-offs from a 50-foot hover, which is not to suggest that you should try it.

The Movies

Film work tends to be done by experienced pilots with whom directors, and particularly cameramen, are comfortable. It is not impossible for the local guys to do it if they are unavailable, but there is a big difference between just circling a target and chasing a car backwards down a ravine - experienced pilots should also have the maturity to keep out of danger by not pushing the machine past its limits, or not get concerned when things go awry.

You will mostly come in contact with the production department, who are responsible for the correct scheduling and availability of equipment, including the helicopter, which gives you around the same status as a typewriter, or sometimes less. You could be on standby all day, and not fly at all, or be told the evening before you will be taking off very early. In short, there will be a lot of jerking around, which should not be taken personally. Ever since an episode in The Twilight Zone, where some people got killed in an accident involving a helicopter on the set, everyone is paranoid about making mistakes and being blamed for them. Very often, the pilot is the only one without an alibi, so that is where the blame will end up. It's all part of the job, and making movies is a very high pressure business (in the above case, the film technicians or engineers, I forget which, were fond of having a beer in the back of the machine, and everyone thought the cans belonged to the pilot).

Cameras are fitted in various ways, typically sidewards-facing, but the Spacecam is a gyro-stabilised affair that lives in a large round casing on the front of an AStar (OK, so it's a Twinstar in the picture - it fits both. We used it for K-19). The electronics are so good that you can take a fair amount of turbulence without the camera even seeing it.

The flying includes a little formation work here and there, some precision hovering, or cross-controlling when you have to make sharp, level turns. You need to be smooth on the controls, operate safely and ensure there is plenty of communication, but most important is knowing the performance limits of your machine. Directions, when given, are in relation to the camera, as in "Camera Right" or "Camera Left". Otherwise, if you don't have an aviation liaison person, you can expect all kinds.

AERIAL SURVEY

This is the process of photographing areas of land from varying heights, the results generally being used for map-making. As a result, this takes place at great heights, but it may get exciting and bring you down to 300 feet, depending on the results required. Aerial survey can give good job satisfaction, especially when you can see the results, and the target appears in every frame as requested by the surveyor.

When doing low-level work, you will be given a large-scale map with flight patterns marked on it, and you do everything by pure map reading. The pattern can be star-shaped, with sets of two or three parallel runs at angles to

each other over the target. The equipment used is something like the Zeiss trilens, which will take one flat and two oblique photographs at the same time. You can work at higher levels with a 35mm, but you will need a navigation aid, like GPS, as close map-reading is not so easy up there.

With 35mm at least, as the focal length decreases, depth perception increases, and required altitude decreases. Your camera can either be aligned transversely (i.e. in landscape mode), or longitudinally; the former makes for easier navigation, and should be used for overlapping strips, but the latter is more flexible. As to results, a 28-times enlargement is used for display purposes, otherwise a contact strip with a 7x stereoscope is good enough for most work except map revision, which needs to be blown up around 6.3 times. A larger scale makes the results easier to read, but you need to fly lower and take more photos.

You need to know such things to use certain tables that give you altitudes to fly to get proper coverage, from 500 to 11000 feet, that is, the lowest for low flying and the highest without oxygen. They also give you the speeds to be flown and the intervals between pictures, which will ultimately tell you how long you will be flying and how much to charge the customer. The book with all these in is called *Parameters and Intervals for 35mm Aerial Photography.*

Normally use a shutter speed of 1/500th of a second, or 1/250th if the light is bad. Make sure the camera is set to the ASA rating of the film. Use a 28 mm lens above 4000' and 35 mm for lower, and focus to infinity. A yellow filter is needed for winter B/W photography.

AIR TESTING

If you're a junior pilot, you may find yourself doing quite a bit of this, anything from just engine running upwards. The reason why junior pilots tend do them is because they're boring and regarded as a waste of time by anyone except engineers. Nevertheless, Air Testing demands your full concentration and everyone due to fly in the aircraft later deserves it as well. One point to bear in mind is that the aircraft is only technically serviceable for the air test - if in doubt, ask an engineer to go with you; if he won't fly, be suspicious!

The least taxing are straight engine runs. When a sliver of metal is detected in oil, there can follow an engine run for anything up to two hours or so (I have known one for five) to see if it happens again. Then there are compass swings where you place the aircraft on a series of headings on an isolated spot well away from large hangars and other machines, while someone with a landing compass stands outside in the cold and rain taking readings. Comparison of your readings with his, adjusted with certain formulae, give the corrections.

The shorter air tests tend to concern themselves with the proper rigging of flying controls. The longer ones creep into the full-blown C of A air tests which are Extremely Official and done under strict procedures. For these, you must be on the Maintenance Contractor's approved list of test pilots, which means having experience on type and flying accurately.

The basic idea is to perform a series of prescribed manoeuvres (timed climbs, for instance) while an engineer takes notes of temperatures and pressures, etc. The results are plotted on performance graphs (by you) so they can be compared against the standard figures in the Flight Manual, which is where you see how accurate your flying really is, when the plotted points end up all over the place instead of being in a straight line. Before you start, though, be sure that the rotors are as clean as you can get them, because their state will make a surprising difference to the climb figures.

SEISMIC SUPPORT

An oil or seismic company operating out in the field needs a helicopter for various reasons. First of all, there are not likely to be any roads, or, at least, no more than forest access roads to the staging area, and people (such as slasher teams and drillers) will need to be moved, as well as their supplies, which will be anything from fuel for the drills to explosives. This will mean a lot of slinging into tiny areas at the end of a longline - in the latter stages, you might have a carousel at the end holding six bags which you must drop carefully in precise locations, as they hold about $6000 worth of equipment each, but a separate company will likely deliver these after you've done the front-end stuff. The expected rate for "production longlining", as it's called, is between 35-45 bags per hour. Sometimes, you will have a Dynanav or Kodiak machine to help, which produces a series of squares on a screen, and when they all line up, you will be on target (although there is a danger here of not looking where you're going when concentrating on the machine). With this taking the strain, the slashers only have to clear a couple of trees here and there.

In the early stages of the front-end operation, the slashers (big guys with chainsaws) will create the helipads so you can position them in every morning for the rest of the week while they cut lines a metre wide for the surveyors to mark out for drillers, who make holes for explosive charges (when the whole lot is blown up, the vibrations are recorded and analysed in the hope of finding oil or gas - alternatives are electric vibrators or falling weights). You will need to know how to work a GPS, as there is some precision involved, although, outside of winter, decent map reading skills are good enough once you know where the pads are. However, when flying the lines so the surveyors can check on how the slashers are doing, you want one that can show you the lines to be flown.

Seismic support is a risky business - in 2004, it had the highest fatal accident rate by a wide margin. However, risks can be minimised by planning and correct analysis of the job in hand, plus refusing to buy into the customer's impatience. Risks can arise from the nature of the area, its altitude, climate, the type of load, helicopter performance and whether Search and Rescue is available.

I'm not a believer in too many meetings myself, but a safety meeting before you start on seismic is essential. Most of the work force will not be renowned for their intelligence and you can expect to give them a considerable amount of education. There will normally be a daily safety meeting which you should attend, and be ready to answer questions on the helicopter's requirements.

Normal requirements for equipment and passengers apply (read this in conjunction with *External Slung Loads*), but the helicopter should definitely have external mirrors, a manual and electrical release for the hook, a fuel low warning light (very important!), and remote gauges. Some machines might have a cable between the skids and the fuselage to stop the cable getting caught up in them.

In calculating an OEI hover capability, forecast winds should be ignored when under 10 kts, and you should only allow half of any above that anyway. However, winds that are too strong will affect the load's flying characteristics:

> *"Had the opportunity to work with some real bone headed customers once - anyway, it was blowing a gale just after we had moved into a new campsite, raining also. The camp boss says its time to sling the gear off the beach up to the campsite, around 1/4 of a mile. My suggestion of waiting for the wind to die down falls on deaf ears and the comment to me was: "If you don't want to sling it up here you can go down and start carrying it up here". This after bending over backwards for these jerks for the*

last 30 days, so I told him if things go bad the load will be punched. His reply: "Whatever."

Well I tell them the loads will have to be light as the pick up point is upwind of the drop off site, and short of hovering backwards for a 1/4 mile, turning downwind will be some fun. Well, two loads go as reasonable as one can imagine with the wind, rain and pilot with a bad attitude - now for the last load - and its just a beauty - plywood on the bottom and assorted junk on top. Unknown to me at the time was my own personal toolboxes (2).

Well, getting it up was a bear, with maybe 10 torque to spare. I try to hover sideways with it...no luck...drags me to the ground....going all the way backwards was a good way to run into something near the camp, so let's try flying this crap and doing a mile wide turn. She gets going okay and into translation okay - up to 30 mph - okay - start to do the wide turn okay now going downwind like a bat out of hell, trying to push enough cyclic to keep her flying. No way! The airspeed hits 0 and we start to sink - the torque is at 100%, so as calm as one can be at this moment in time, I know that this load is not long for the world. Try turning into wind, no way will she turn - heading for the ground - left hand ready for the emergency release just in case the electric one fails, going to wait for the last minute before punching it off - about 10 feet from the ground - bombs away, gain control of old Betsy again, calmly park her near the camp, shut down, blades are still turning when the camp boss and his cohorts come over and now are saying that I did that on purpose and want to kick the crap out of me.

Well, seeing as they outnumbered me six to one, I calmly announced that if they should like to join into fisticuffs, get in line! As I drew a line in the dirt, I said I would gladly oblige them - one at a time. After no one took up my offer and I tried explaining what I had told them before this dog and pony show got started, they shuffled off and I went over to see what had become of the bomb load. To my displeasure I found what was left of my toolboxes....flattened to about six inches high....

Doug Potts

Once you have calculated HOGE performance for your density altitude, whether single- or multi-engined, the load should then be reduced by a further 10%. A ceiling of 600 feet and 3 nm visibility should be regarded as the minimum weather conditions.

Flight Following

There will be a truck acting as a flight watch station and you will be expected to report in every time you land and take off at any helipad.

The person in the truck (usually the medic) will be keeping a log of all movements and radio calls and will therefore have the most information to hand if an incident occurs. Expect also to be given an Emergency Response Plan, which is a bit of paper telling you what action to take in emergencies, together with this information:

- Your location (Lat/Long)
- Who's in charge
- Any Radio frequencies
- Police, Fire, Ambulance, Hospital
- Other helicopter companies
- Medic
- Safety etc.

All this concerns your second job of aerial ambulance. You should always have enough fuel for the nearest hospital, but not so much that you can't lift the patients, which, typically, will be the heaviest guys coming out of the tightest clearing with the tallest trees.

You should leave a map with landing sites and potential routes behind with the flight watch station, so they don't waste time in the wrong places if they suspect you have gone missing.

Tip: Try to get fuel in the staging area, to save dead flying time and unnecessary starting of the engine. This is not always practical, but as it saves them a lot of money, you should find them more than interested. More importantly, it stops you being away getting fuel when an accident happens. A handheld radio is also useful, for keeping in touch with the ground crews when the helicopter is shut down, and for you to call for help with when you see a bear tying a large bib round its neck.

Finally, bear in mind that many passengers will be competent professionals in their own field and will be used to helicopters, probably flying with several companies aside from yours. All they will be interested in is safe transport, so give them a smooth flight and a good briefing. Word gets around.

The Township System

When you're working on the oil patch, you will typically be asked to go to a lease, with coordinates looking something like this:

```
SW-1-87-18-4W
```

Unfortunately, they don't always work in Lat & Long, but use the township system, so you have to convert between them. Also unfortunately, there are only a couple of out-of-date DOS programs that do this, or maybe the odd website, but you can't always use the Internet or carry a laptop. Equally unfortunately, there are few PDAs left that run DOS, except the HP 100 (though the program works with a DOS emulator on a Psion 5mx or a Pocket PC). This bit is to give you an idea of what goes on so you have half a chance of working it out yourself (although it specifically refers to Alberta, the principles are good for the rest of North America).

Land is West of the 4th, 5th or 6th Meridians (110°, 114° and 118° W, respectively). Between the meridians are six-mile-wide columns called Ranges which are numbered consecutively from East to West starting at Range 1 West of each meridian. Townships are six-mile-wide rows that intersect ranges and are numbered consecutively from Township 1 at the Montana border to Township 126 at the Northwest Territories border (*township* also describes the six by six-mile square formed by the intersection of ranges and townships). Townships are numbered northward, starting from 1 at the 49th parallel up to 126 at the 60th parallel (the north boundary of British Columbia, Alberta, Saskatchewan, and Manitoba). For example, Lethbridge is roughly at township 8, Red Deer at 28, Athabasca about 66, and Fort McMurray about 89. Townships are divided into 36 sections, each section measuring one mile by one mile. Sections can then be divided into quarters (NE, NW, SE and SW).

The trouble is, meridians converge as they go northward. For example, the distance between the 4th and 5th Meridians along the 49th parallel is about 182 miles (293 km) - at the 60th parallel, it is about 139 miles (224 km). There are 30 ranges between the 4th and 5th meridians along the 49th parallel, but convergence reduces this to only about 23 along the north boundary of Alberta.

Base lines run between the Initial Meridians. A base line approximates a latitude circle from which townships are projected north and south to the *correction lines* (see below). Base lines are four townships apart. The USA border is the first base line; the 2nd lies between townships 4 and 5, the 3rd between 8 and 9, and so on northerly in regular order. For example, the 14th base line (between townships 52 and 53) runs along part of Jasper Avenue in Edmonton, and the 24th base line (between townships 92 and 93) runs north of Fort McMurray.

13	14	15	16	
12	11	10	9	
5	6	7	8	
4	3	2	1	C D / B A

Correction lines are east-west lines, midway between base lines, on which *jogs* provide for convergence. They are also four townships apart. The north boundary of Alberta is about the 32nd correction line. The jogs along a correction line increase in length as you go west from an Initial Meridian. Each section is one mile on a side, and contains 16 *legal subdivisions* of 40 acres each (see left). These smaller tracts are used for smaller divisions of land bordering on rivers and lakes, Indian reserves, settlements, and for oil and gas well spacing units.

In the example above, the first figure (1) is the section, the second (87) the township, the third (18) the range and the fourth the Meridian.

```
SW-1-87-18-4W
```

SW refers to the corner of the section.

AVALANCHE CONTROL

This involves applying explosive charges to selected places in an avalanche start zone, so you need permission to carry dangerous goods. You must be precise, and there will be a qualified Blaster (Bombardier) on board to dispense the explosives, who will have a body harness secured to the helicopter in two places - it must *not* have a quick-release mechanism. The doors will be off, so you need to dress properly and watch your speeds and C of G, especially lateral. In fact, the weight of explosives and people required for the job must not be more than 75% of the useful payload. Ski baskets, or any other restrictions to dropping must be removed beforehand.

The area should naturally be closed. Primers must be prepared before entering the aircraft and pull-wires carried separately - none are to be adjusted inside, and they are to remain in the same container at all times. Only the explosives to be used on a particular sector should be carried, and they must be pushed away from the bombardier's seat position where you can see them, not above, behind or below. You have 90 seconds after dropping the first bomb to do the rest before you must pull away to a safe area for observation. If there is no assistant, that is, just you and the bombardier are on board, you can only drop three charges anyway.

SNOW N' STUFF

Dry snow and wet snow are different, and so are dry and wet avalanches. In simple terms, the former are caused by *overloading* the strength of buried weak layers, and the latter by *decreasing* the strength of buried weak layers.

Wet snow that has been melted and refrozen several times is also called Corn Snow, which has a crust on its surface that will support the weight of your helicopter when frozen, but turns to deep slush during the day, because the water dissolves the bonds between the crystals. However, the surface tension of water also acts as a powerful glue, which is why you don't get avalanches all the time - only when saturation occurs do the bonds dissolve.

AERIAL HARVESTING

This is removing tree crowns or cones from standing trees with a helicopter, for monitoring the progress of growth, insect or disease infestation, and pesticide or fertiliser trials. There are two ways of doing it. The first involves an unmanned device underneath the machine which uses the downwash to separate foliage from the tree, which ends up in the device. A Branch Collector (from the Branch Office ☺) cuts branches from the main stem, a Rake collects cones by stripping them from the foliage with stationary or moving tines, and a Top Collector takes away the upper portion of the tree crown. The second method is really aerial clipping, since it is done by someone from the rear door with an battery-powered chainsaw while you hover near the top of the tree. Whatever is cut ends up in the helicopter, so there is a barrier between the two parts of the cabin to make sure you don't get a bump on the head (believe it or not, this was originally done with a steel box with retractable doors, the edges of which cut the tree as the box was lifted up from it, but it proved to be slow and cumbersome, and strained the equipment). As well as the clipper, a navigator tells you where to fly, since you will be busy enough at that level. Don't expect to fly more than 6 hours a day, or in wind over 15 kts. The aircraft should be on low skids, with no bear paws, or items that can get caught. You should be an ace at longlining.

You start off with a hover somewhere between 25-50 feet above the trees concerned, using up to 85% torque, so you've got a reserve. Keep them between your right shoulder and just forward of the front edge of the front door. Then descend to within eye level of the target, always being aware that trees sway in the wind and you can either get hit by another or hit something else as you try and keep up with the tree you are working with (your downwash won't be helping). Its top should be level with your eyes or the cabin roof, slightly forward of your right shoulder. Don't fixate on the tree top, but something else that isn't moving. Slide the skid in between the branches until the stem of the tree is against it, using a little collective just before contact. Then lower the collective so the skid rests against the lower branches to keep it steady for the clipper, who shouldn't take more than a minute to do what's required. Before you move away, check the skid isn't caught on anything, then increase collective and use left cyclic to pull you away gently. Don't move directly sideways, as slanting branches near the tops of trees could snag you.

A couple of points - if you bend a tree beyond the cyclic stops, when it rebounds you will be out of limits if you try to correct things. You will do best to keep level and use left cyclic with the rebound. Don't go straight up or you might get dynamic rollover. Trees with bent tops indicate heavy crops and should be avoided, because they will have too much momentum. If you have to work round a tree top, hover away and assume a new position every time rather than spinning round it.

WILDLIFE CAPTURE

Normally done with a pilot and gunner (who has a net gun), but there may also be an assistant (called a mugger) as well. If there isn't, and you have to do the job instead, you should only handle the head of the animal, so you don't get injured unnecessarily. Your briefing should include getting in and out of the helicopter whilst in the hover. A slightly forward C of G is preferable. Keeping the rotor disc as flat as possible, and avoiding tight turns, approach the animal from behind, slightly away from the skid and forward of the gunner, almost definitely with a sideslip. Don't fixate on the animal, but keep it in view with peripheral vision. The reason for the flat disc is to keep it out of the line of fire, and high-G turns will upset the gunner's aim. Sometimes the wildlife will be seals, so you have to wear lifejackets and stuff, and dodge the waves when they get a bit big (sometimes you have to use the waves to get you back in the air again).

FROST CONTROL

Note: This is based on procedures that were generously circulated by Peter & Margaret Garden of Peter Garden Helicopters, previously of Gore, Southland, NZ. Other bits have been suggested by other pilots.

A helicopter can raise the air temperature in a particular area by pulling down warmer air from an inversion layer above or by keeping the air moving to stop cold air from pooling in hollows, etc. You use the OAT gauge to locate the inversion level, then establish a hover height from where to drag down the warmer air. One farmer in Washington State grows 1000 acres of cherries and uses a 212 to dry them with, since water collecting near the stem will cause them to split. The downwash is increased by loading the helicopter up with sandbags.

Hazards

There are a number of these:

- Firstly, it is done in low light conditions, if not darkness, so you need visual reference, and awareness of normal optical illusions. The area should also be inspected immediately beforehand, in daylight, which includes positioning the aircraft. Obstacles can then be noted and illuminated.

- As the hours are unsocial, be aware of fatigue.

- All lights should be on, with instrument lights operational. The hook mirror should be removed (as should anything else that might cause reflections) and you need a torch in the cockpit. Warn ground staff and others about the dangers of shining lights that may affect your night vision.

Preparations

- Ensure that equipment required is on board, and equipment not required is removed. Do a thorough preflight (including lights!) and refuel as required. Ensure that you have overnight kit and that a suitable bed is available to provide adequate rest beforehand.

- File a flight plan with Base or ATC, and make sure that you and the client know where you will be landing (as close to the scene as possible so you can hover taxy to the area).

- Once on scene, check the area for obstructions, especially wires or cables, including those nearby and where you will be hovering.

- Cover the windscreen to keep condensation at bay. Maybe keep the cockpit heated too.

Operational Procedures

- To calculate the latest time you can start, add the time to get the aircraft and crew prepared, the ferry time and site inspection time together. Deduct it from Evening Civil Twilight to establish the final callout time

- Once advised that temperatures are close to minimum, place safety lights as necessary to provide warning of obstructions and delineate block boundaries

- Preflight the aircraft and establish the maximum flight time for the fuel on board (allowing for minimum fuel). Note the time that you will need to refuel

- Remove the covers and start the aircraft

- Check that the OAT gauge is operational and visible

- Check communications with client/ground crew is operational

- Ensure that condensation inside and outside the bubble has dried off

- Hover taxi to the operational area, making sure you don't lose sight of the ground and your bearings. Once on site, check the OAT and establish how high the inversion layer is. This can be extremely dangerous and you MUST remain in sight of the ground! If you cannot establish the inversion layer, you will have to establish the low point of the area where the cold air will pond and, with the help of ambient air flow, keep the air moving through it

- Continue operations until the air temperature in the block is raised enough or the client advises you to stop

- Once operations have ceased, collect the equipment (refuelling gear, hazard lights, etc.). If this occurs before daylight, tie the aircraft down and refit the covers until daylight arrives so you can fly home

Responsibilities

OPERATIONS MANAGER

- That clients are aware of their responsibilities, including giving enough warning of the requirements, advising of hazardous obstacles, providing pilot and crew with appropriate rest facilities, etc.

- That properly trained and approved pilots are used, who are fully briefed

- That all necessary resources are available, including hazard lights, OAT indicators, and fuel supplies

- That appropriate documentation is completed pre- and post operation, including flight plans and flight following

PILOT

- That required equipment is on board before departure

- That training is current and that appropriate authorities are held

- That twilight requirements and location are understood

- That required documentation is completed, including flight plans

- That client and ground crews are briefed

- That a fuel plan has been formulated

- That safety hazard lights are placed in those areas, identified in daylight, that pose a risk or potential risk to the safety of the operation

GROUND CREW

- That they have been fully briefed on the operation, including the location

- That they carry enough fuel for the operation and the return flight

- That they carry necessary equipment for the operation

- That, where possible, they carry out, along with the pilot, a thorough inspection of the block in daylight before starting operations

- That, during operations, they maintain a continuous radio watch on the appropriate frequencies and advise the pilot of potential threats to safety

CLIENT

- That there is enough notification of pending operations to allow time to prepare and dispatch a helicopter to arrive in time for inspection in daylight.

- That they provide an area to park the helicopter near the block so the helicopter can hover taxi in ground effect to the block

- That all hazards or potential hazards are pointed out to the pilot and a method of illuminating them during the operation is established

- That suitable rest facilities for the pilot are provided

- Were possible, provide continuous support for the pilot and ground crew

EXTERNAL SLUNG LOADS

4

Helicopters can go where cranes would be impractical or more expensive, or you might not be able to get a particular load inside the machine (maybe you wouldn't want it there anyway, if it's a dead animal or explosives). Another reason for slinging might be that there's nowhere to land in the first place!

In theory, you can lift anything, provided the payload is available; I've even been asked to quote for lowering 800 feet of unrolled telephone cable down a mine shaft, because the drum it was rolled onto wouldn't take the weight. However, more common tasks are logging*, placing air conditioning or ventilation gear on the roofs of tall buildings, erecting towers (where inches count), pulling cows out of bogs, picking up water to put out forest fires (water bucketing), dropping solution over forests (top dressing) or moving seismic equipment about. In fact, many tasks done with a helicopter are really extensions of load slinging, and, in remote areas, this will be a major part of your bread and butter - really specialised stuff will be found as subheadings below. A typical line length is 25 feet, but can be up to 200, so don't forget to include the line as part of the payload - it will be heavy (try about a pound a foot). Ex-military pilots will have done all this as part of their original flying course, but, in some companies, "training" is once around the circuit!

Logging, officially, is removing logs from areas where *all* trees have been felled. It is very fast, with lots happening at once, and there will be a smaller helicopter to carry the used chokers every 75 minutes or so (chokers are lanyards with the equivalent of a slip knot which tightens as the load is taken up, making it more secure- see later). It is not an operation based on finesse, as the machines are continually using full power cycles and undergo a lot of twisting, etc. If you're buying a helicopter that was used for logging, inspect it *very* carefully! *Selective logging*, on the other hand, is removing wood from where trees are still standing, and is considerably more dangerous, at least to workers on the ground, because the downwash could dislodge all sorts of stuff. *Cedar salvage* involves moving loads of cut cedar blocks, which should all be of a similar length for best stability and uniform lifting weight, as the turns are not weighed every time (a *turn* is a load of logs). Logs will be taken from high ground first so there is less risk of anything falling on workers below and you can see

what's going on better. They will be delivered to holding points on land or in water.

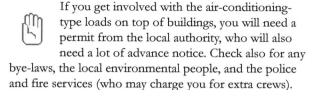

If you get involved with the air-conditioning-type loads on top of buildings, you will need a permit from the local authority, who will also need a lot of advance notice. Check also for any bye-laws, the local environmental people, and the police and fire services (who may charge you for extra crews).

Passengers should not normally be carried when slinging, but some authorities, such as Ontario, insist upon you carrying an observer. Even though they are not officially "passengers", normal passenger facilities and briefings must be available. Having said that, adding some extra weight in the front seats, especially in a 206, helps you see the load and landing site a bit better, as the nose will drop slightly.

Good line flying starts with good airmanship - just to make life interesting, when logging, you might also be ducking and weaving around another 2 or 3 support ships, a couple of 214Bs and the odd 61 or Crane. *Do not* get in front of a big ship, whether or not it has a turn on. Keep focussed on the basics, e.g. smooth power application and reduction, with light feet on the pedals. Also, review getting out of vortex ring. Logging and seismic drill sites are never where you'd like them to be, and you will work a lot in downflow. In fact, most slung loads are only carried a short distance. If you are doing several hundred approaches a day, doing each one is into wind takes a lot of time, so sometimes you will have the nose into wind for the delivery, with a tailwind for the pickup, or even a crosswind at both ends, so the challenge (for the production longliner, at least) is to learn how to move quickly with unfavourable wind conditions, which will ultimately depend on how accurate you need to be and the power you have available.

There are a range of "skills" amongst helicopter pilots, but one of the most important of those "skills," is respecting your equipment and fellow pilots. This has precious little to do with actual hands and feet, and everything to with common sense and the aforementioned respect.

Any machine will only lift what it will lift within limits. This of course fluctuates with altitude, temperature, wind speed/direction, etc etc. You will come to learn that most helicopters will lift more than what they are certified for, and that "lifting" can sometimes be a whole lot easier than "putting down." Certain types will lift large amounts more than ever intended by the manufacturer.

There is a huge difference between barely moving a load, and moving one safely and effectively, with finesse, and respect for the machine and those who will fly it afterwards.

ADMIN

In UK, you will find it useful to check out the following:

- **CAP 426** - Helicopter External Load Operations

- **ANO** - Public transport aircraft & suspended loads, Picking up and raising of persons & articles, Dropping of articles & animals

- **Rules Of The Air** - Marshalling signals

In the USA, check out **Part 133** of FARs, and **Section 702** of CARs in Canada. In those countries and Europe (under EASA), external loads come in four classes:

- **Class A** - the load cannot move freely, or be jettisoned, and does not extend below the landing gear. In other words, it is usually bolted to the aircraft (e.g. a stretcher, spray kit or fuel tank)

- **Class B** - the load is jettisonable and is not in contact with land, water or any other surface

- **Class C** - the load is jettisonable but remains in contact with the surface (like when towing, or wire-stringing, cable-laying, etc.)

- **Class D** - the load is not one of the above and is specifically approved. It usually includes a person, for which you mostly need two methods of release and a multi-engined helicopter that can hover on one engine in the prevailing conditions, with appropriate engine isolation

Most loads will be Class B, and this is what the following pages will concentrate on (for longline rescue (Class D), see *Search & Rescue/Hoisting*).

There may be a little paperwork to do first - your customers will likely need to be made aware that cargo insurance is available, if your company provides it, and authorise the flight by signing a damage and injury waiver agreement. A couple of other points: Your C of G will be fairly near its ideal position with a load on, but maybe not when you release it, especially if you're low on fuel to lift it. Also, loads that must be guided into place or secured while attached to the helicopter must be given special consideration, especially when briefing people on the ground - they must NOT go anywhere underneath the load or be in any similar position that would be dangerous if the load gets released. Although you may not be responsible for the accident, because a helicopter is involved there will be no end of paperwork, and no doubt for more than one authority.

When organising a job, it is often better (though customers do not always see why) to take smaller loads, which not only means that you have more power in hand, but can also take more fuel and stay airborne longer, and probably shift more loads per hour because you're not stopping to top up all the time. Even though the total flying might be slightly longer, this could mean the difference between getting the whole job done in a day or staying overnight and finishing the next day. Unfortunately, people don't always see that sort of big picture. They consider that they are paying for the complete payload on each lift and expect to use it in full every time.

GROUND CREWS AND EQUIPMENT

Without a mirror, the ideal team on the ground consists of at least three handlers at every point of pickup or deposit, so, in a simple lift from A to B, you need six, although this could be reduced with decent communications. All procedures given here are based on the assumption that they are not available, but things will go so much better if they are - just make sure that any instructions given don't require acknowledgment, as you will not only have your hands full, but it's also easy to hit the load release button when moving your hand to transmit, and you won't be able to hear them properly anyway (actually, that goes both ways; very often both hands are needed by the loaders, so they put their radios in their pockets, and can't hear you, which is why having them in helmets is recommended).

Otherwise, ground crews should also have hard hats (maybe different colours for different groups) with chin straps, goggles or safety glasses, protective gloves and a metal probe for discharging static electricity. As mentioned above, radios in the helmets are most useful.

One person (of the 3 at each point) would be for marshalling and the remainder for hooking up, etc., but, in remote areas, you could well be operating by yourself, including picking up the load, which may mean continual shutting down, etc. Expense is not actually the reason, although it helps; in the Arctic, for example, you don't leave somebody by themselves in case you can't get back, but, in general, you are dropping stuff off where you can't put people anyway. It's not as hazardous as it sounds - for takeoff, you just need to be far enough behind the load to stretch the line properly, with no kinks, and make sure it's straight, so it's away from the landing gear when you lift into the hover.

Always lay the load out in front!

 However, don't attempt anything alone without a mirror

Note: Some helicopters allow a higher all-up weight when slinging. The maximum hook load (as in the Flight Manual) has nothing to do with payload, but is merely the weight the hook *on its own* can stand as a structural limit. If you try to lift the max hook load in a 206, there will be no room for you! In other words, the weight of the *helicopter* and *not the capacity of the hook* determines the payload (and that comes from the HOGE graph).

Ground Equipment

Radios and whistles (for communications) should be included, plus weighing scales, accurate to at least 25 lbs, and capable of weighing more than the maximum payload. Items should be labelled to show their safe working limits (if relevant) and the dates of any inspections, which should be carried out regularly, as should maintenance. If labelling is impractical, you could always use coloured paint.

Static Electricity

This comes from a number of sources, the main ones being engine and precipitation charging from friction between the aircraft's surfaces and airborne particles (there is also a risk when thunderstorms and snow particles are about).

Although the capacitance associated with this is small, voltages as high as tens of kilovolts can throw people to the ground, as well as being dangerous near potentially explosive cargoes or fuel tanks, or even short-circuiting the hook's electrics. It's for this reason that an earthed static discharge probe is applied to the hook *before any contact takes place* and the procedure kept up as much as possible. If you can't get one, at least make the guys wear thick rubber gloves.

Tip: Transmitting on HF or VHF can discharge a great deal of static through the antenna.

Helicopter Condition

Depending on the job, the rear doors may need to be removed so that used harnesses can be placed inside quickly from either side - very often dropoff points are in places where you can't land, but only come to a very low hover. Also, there is a little less weight for the machine to carry (rear doors weigh 50 lbs on a 206). If you're doing vertical reference (see later), you will need your front door off anyway so you can stick your head out of the side. However, taking any door off will mean checking weight and balance and performance figures, and your V_{NE}, as there may be a speed restriction.

 The AS 350 should have *all* doors off on each side!

You need a mirror (or two) so you can see the behaviour of the hook and load (especially for positioning if the doors are on):

Hooks must be enclosed, that is, they should not allow the load to come out when you are flying. They must be checked for consistent operation, as must all standby release methods - don't accept the fact that the solenoid clicks as evidence of it working. If there's no-one else around, put a rope in and pull on it when you operate the mechanism. After you operate the manual release, check that the Bowden cable between the hook and the body of the helicopter doesn't bind and stop the hook from rearming. Witness marks should be aligned on the knurled knob or lever and the hook body, make sure the hook moves over its full range of travel and that the bungee cord keeps it tight against the bottom of the fuselage, so you don't land on a vertically extended hook, which may get snagged (garbage on the landing site has been known to pull the manual release enough to allow the load to work itself free).

Sling Equipment

Ropes, lines and hooks are part of the helicopter and should be treated as such, as a helicopter without them cannot earn money. Because of the direct connection to the aircraft and the potential for damage, sling equipment should only be used on helicopter operations, and worn or frayed items should be discarded. You're generally allowed up to 10 randomly distributed frayed wires on a steel sling, or 5 in one strand, but you would be best off with none at all (fuzzy broken strands on a rope actually protect the ones underneath, but 25% is the maximum for braided ropes - 10% for twisted ones). Nylon deteriorates when exposed to petroleum, and wire rope rusts and doesn't like

being mistreated, so protect them from moisture and heat, and inspect them regularly. The maximum length for nylon or poly rope should be 6 feet.

There should be as many ropes, strops, nets and hooks that can be made available, as more will always be required than you think. Steel slings are best, though ordinary rope will do, provided it doesn't have a tendency to stretch or bounce up if it breaks (for this reason, don't use nylon lanyards). At the very least, you need one set of slings at each drop-off point, so while the first load is being undone you can be on your way back with one and not waste flying time (when logging, a smaller helicopter is used for this job).

Equipment should be able to withstand 6 times the anticipated load (a ratio of 6:1) because flight conditions may increase its weight artificially, and to allow for deterioration (abrasion, cuts by the load, etc.), errors in load weights, and the effect of the sling angle (later). The correct term is *Design Breaking Strength*.

Chokers, technically, are lanyards that hold loads with a slipping noose which tightens under tension - the harder the pull, the tighter the grip (left). Use chokers at the end of (i.e. attached to) another line rather than by themselves so that, if they break, they are kept away from the blades by the other line (this also helps prevent *collective bounce* - see later). If a sharp bend is necessary, you can use a short strap around the load instead, and attach it to a straight lifting line with a clevis.

Hooks

Long lines have an electrically operated hook at the end, which makes them an extended hook, so there is also an electrical cable for the release mechanism, which should be carefully looked after; that is, the line and cable must be detached properly by ground crew, otherwise you will strip the plugs, which won't impress the engineers (if the insulation comes off, you may also get a short circuit).

If the hook has a keeper, have it locked so that the size of the ring attached to it does not restrict you.

Use barrel hooks on a sling for lifting fuel drums, although you might find varying designs to suit different helicopters. For example, a 206 might have one for 2 drums while a 205 might have one for 5. Whatever you use, use the bungee cord to keep them together when there is no tension. Again, don't connect the hooks directly to the aircraft, but to a sling.

Mechanical hooks can be found at the end of lanyards, slings and swivels, and made of carbon or steel alloy (carbon ones are bigger, for the same weight rating).

The picture below shows two variations - the one on the right has a swivel. Neither will necessarily have a weight rating, so the scaffold hook may well be used instead (bottom picture)

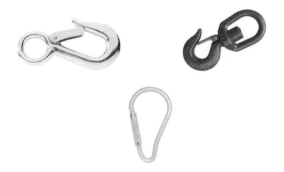

PEAR-SHAPED SLING RINGS

The pear ring is the most common item used directly with the hook, often with a shackle (below) if it is too small.

Its size must match the hook, as rings that are too large may flip over the edge and unhook. Rings that are too small may hang up and not release properly.

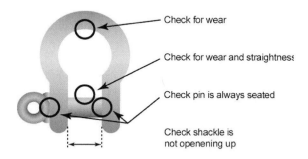

½" pear rings are commonly used on top of ⅜" long lines on light aircraft, and ⅝" rings might be used on a ½" line with a Bell 212.

Inspect for distortion, extreme wear and severe corrosion.

SHACKLES

Shackles are coupling links used for joining ropes and chain together or to another fitting, usually made of wrought iron or mild steel. U-shaped ones are called *straight (or Dee or Chain) shackles* (right, below) and are used for permanent connections. Those with curved sides are *bow, or anchor, shackles* (left, below). Bow shackles are weaker than straight shackles.

Note: Some texts say the Dee shackle is a Clevis - it isn't!

The ends of a shackle are called the *lugs*, the space between them is the *jaw* and the part opposite the jaw is the *crown*. The inside width or length is the *clear* - hence the phrase *long in the clear*. The jaw is closed by a removable pin that passes through a hole in each lug, and a general purpose shackle is usually named after the way the bolt is secured in place. For example, the screw shackle has the end of the pin screwed into one of its lugs. The pin should be lock wired, after being screwed in hand tight.

Inspect for deformation, wear and corrosion, as shown.

- Check for wear
- Check for wear and straightness
- Check pin is always seated
- Check shackle is not opening up

When using a shackle with a rope, using the shackle downwards can allow the rope to slide from side to side and strain one side or the other. If a shackle is allowed to be pulled at an angle, the legs may open up.

If it rolls fore and aft, it could also undo the pin, unless it is lock wired:

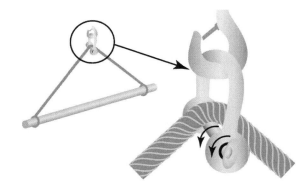

You could place the screw-pin in the hook, padding the empty space either side with washers, and use it upside down.

Packings   Hook

LINK RINGS

These are used with barrel slings, and those with multiple legs.

Inspect for distortion, wear and corrosion.

SWIVELS

Swivels stop the load from spinning and unravelling the line, and reduce the torsional load on whatever you shove into the hook. That is, you put the swivel in the hook, then attach the load to the swivel (which has a thrust bearing in it). With multiple loads, use multiple swivels, as one load may spin faster than another (lighter or more fragile ones should go above heavier ones). Swivels have a rating stamped on the body. Electrical ones should be treated with special care - they are expensive.

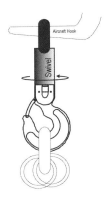

Aircraft Hook

Swivel

Inspect for wear on the attachment point, plus rough bearings and binding during rotation. they need regular regreasing, at least every three years.

Note: Don't put a swivel above a longline! It should go on the remote hook (it also provides extra weight).

Lines

The way to get maximum use out of ropes, etc. is to select them properly in the first place, use them in the right way and retire them in time. A thicker rope will always outlast a thinner one, because the distribution of surface wear is better. A stronger rope will also last longer because the load it supports will take up less of a proportion of its working strength - a 200 lb load is using up only 2% of the strength of a rope with a breaking strength of 10,000 lbs, but it will take up 4% of one at 5,000 lbs, so the latter is working harder and will have to be retired sooner. Braided ropes are stronger than twisted ropes, which is why climbers use them - they have higher strength and stretch less. Their round, smooth construction also distributes any wear over a greater area. Wrap the first 3 metres with high vis tape crisscrossed or braided from the hook down so that it is really visible in the mirror. The new composite longlines have a sheath velcroed shut or zippered shut all the way down which can be dangerous as they have been known to open up and trail behind from the hook end and flail around the tail. They should be tiewrapped for safety every few metres or taped to prevent this from happening.

SHOCKLOADING

An object's ability to return to its original size or shape when external forces are removed is called *elasticity*. For example, a rubber band usually returns to its original shape after being stretched, but if it is overstretched will be unable to. When a substance is on the verge of permanent change, it has reached its *elastic limit*, at which point the substance is said to yield. Put another way, the elastic limit is when the substance (a line, in this case) will no longer return to its original length after a load is removed. Once a steel line has passed the yield point, it stretches faster for the same weight. However, the rubber band's ability to resume its shape also depends on a relatively short time of stretching, so, even if the stretching is within limits, if it's done for a long time, deformation can still take place. It can be affected by the material used, the load applied, duration of stress and the temperature, as warmth can make material more pliable. A sudden increase in strain is called *shock loading*. You get shockloading during slinging when rapid movements (intentional or otherwise) cause the load to exceed its nominal value sharply and suddenly (a gust may lift it slightly, then drop it). This puts a stress on the line.

The *Safe Working Load* (SWL) used to mean the breaking load of a component divided by a safety factor, but it does have legal implications, so the term has been replaced with *Working Load Limit* (WLL) for hooks and slings, etc.

The WLL is the maximum mass or force which a product is authorised to support in general service when the pull is applied in-line, with respect to the centreline of the product. It is specified by the manufacturer. The Breaking Strength is an average loading at which cable samples were found to break under laboratory conditions in straight line pull with a constant and predictable increasing load.

Luckily, even steel cables stretch, which will help to absorb shockloading, but short cables absorb it less.

You can work out a rule of thumb Safe Working Load for wire ropes in tons like this:

$$SWL = Rope\ Diameter^2 \times 8$$

The rope diameter is in inches. This figure may change if you do strange things to the line, like bend it, or use a choker hitch, which reduces the SWL by 25% (the sharper the bend, the greater the reduction). For example, a 1" line will become rated at 13450 lbs instead of 17950 if there is a bend in the rope at the clevis. In any case, the SWL of any piece of equipment should be 20% of the maximum capacity of the hook.

Spectra or Kevlar lines are strong, but light, and will trail after you more than a steel line, so observe the maximum external load speed. These lines also get longer when new, so, if you can't pre-stretch them, allow a good length of extra cable. A **non-rotating** steel line has an outer jacket wound the other way to the inner core so that it resists rotation and unwinding.

Wire rope is steel cable that is made from multiple wires. A single wire rope line is called a single hitch, often also called a lanyard. It is a short wire rope with thimbles, a pear ring on the top and maybe a swivel hook on the bottom. Synthetic lines are popular, but there is no accurate inspection procedure, and no way of knowing how much strength is left in them, so they need to be changed more frequently. There have been very few broken wire-based lines, whereas there have been many broken synthetic ones. However, they are useful for operations near power line, due to their insulation.

³⁄₈" diameter wire rope rated at 2400 lbs SWL might be used with light helicopters, and ½" diameter rated at 4400 lbs with medium helicopters.

When inspecting a line, look for excessive abrasion, glazing (from heat, possibly friction when ropes rub together), inconsistent diameters (flattening, bulges, etc.) or texture, or discolouration. Ultra violet light may degrade Kevlar lines, so they should be kept out of the Sun (although spectra lines fade in colour, their strength seems to remain). A line cover helps with this and keeps the electric cable under control.

High Modulus Polyethylene (HMPE, or Spectra) ropes, from which Amsteel Blue and others are made, begin to suffer progressive strength loss at 65°C, so don't land where the exhaust can affect them. If they get dirty or wet, they will conduct electricity, and dirt will degrade the fibres (look for a white powder-like substance).

Don't beat the hook on the ground or against obstacles (stumps or rocks, etc.), and don't drop a long line from the air. Land and detach it by hand on the ground. If it is dragged through the trees and the aircraft lands on it, the electrical wires will be damaged, if not the line itself. Hooks are quite difficult to repair without tools, which you won't necessarily have with you. You're generally allowed up to 10 randomly distributed frayed wires on a steel sling, or 5 in one strand, but you would be best off with none at all (fuzzy broken strands on a rope actually protect the ones underneath, but 25% is the maximum for braided ropes - 10% for twisted ones). Nylon deteriorates when exposed to petroleum, and wire rope rusts and doesn't like being mistreated, so protect them from moisture and heat, and inspect them regularly. The maximum length for nylon or poly rope should be 6 feet.

Tip: Keep lines coiled in a bucket when not in use.

There should be an eye in each end of a sling, ideally reinforced with a steel thimble, to protect the rope.

If a cable has been doubled back on itself and clamped to make a dual line or shorten one (see below), do not remove the clamp and use the line for another job. The bend is 180°. When straightened out, the inside strands will break and weaken the cable. Knots themselves will not weaken a rope, but the bends involved will.

The shackle pin should be the same thickness as the rope - DON'T use a bolt - it will bend).

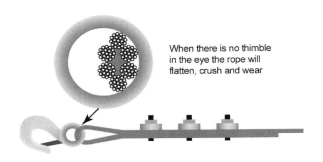

When there is no thimble in the eye the rope will flatten, crush and wear

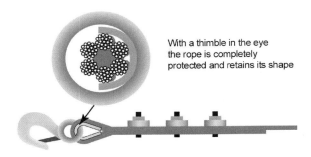

With a thimble in the eye the rope is completely protected and retains its shape

MINIMUM SLING SPECIFICATIONS

	0-1500 lb (B206,AS350)	**1500-3000 lb** (204,205,206L)	**3000-5000 lb** (204,205,212)	**5000-10,000 lb** (S61,AS332)
Lanyards (Steel Core)	7/16"	1/2"	5/8"	3/4"
Lanyards (Poly)	5/8"	1 1/8"	1 1/4"	1 3/8"
Cables, Chokers (6' max)* Single point Multi point	7/16" 3/8"	1/2" 7/16"	5/8" 1/2"	3/4" 5/8"
Shackles and Clevises**	1/2"	5/8"	3/4"	7/8"
Nets	1500 lb	4000 lb	5000 lb	10,000 lb
Extended Hooks	7/16"	1/2"	5/8"	3/4"

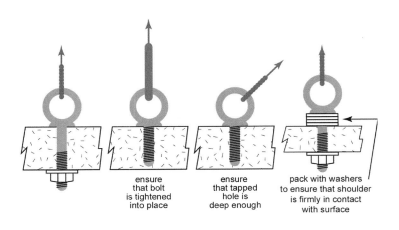

ensure that bolt is tightened into place

ensure that tapped hole is deep enough

pack with washers to ensure that shoulder is firmly in contact with surface

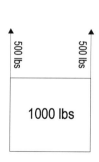
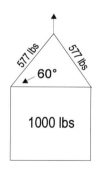
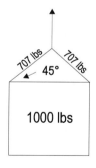
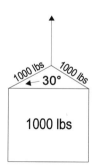

EFFECT OF SLING ANGLE

The strain on each leg of a sling increases as the angle between it and the load decreases:

In the picture above, the sling legs on the left (that is, with a 90° angle) have the least load to carry, whilst the one on the right (a 30° angle) has a strain on each leg equal to the load itself. This is relevant when using the right hand columns of these Safe Working Load tables:

Rope Size Ins	Vertical Hitch (lbs)	Choker/ Hitch (lbs)	Basket Hitch* (lbs)	30° 2 legs (lbs)	45° 2 legs (lbs)	60° 2 legs (lbs)
¼	920	700	1840	1600	1300	920
3/8	2020	1520	4040	3500	2860	2020
½	3740	2790	7470	6490	5270	3740
5/8	5600	4190	11210	9740	7910	5600
¾	8080	6060	16160	14020	11460	8080
7/8	10900	8140	21790	18920	15480	10900
1	14180	10680	28360	24600	20060	14180
1 1/8	16660	12500	33320	28900	23500	16660
1 ¼	20740	15500	41480	36000	29260	20740
1 3/8	25340	19000	50680	43880	35840	25340
1 ½	30620	22960	61240	53040	43300	30620
1 5/8	35900	26920	71800	62180	50760	35900
1 ¾	41170	30680	82330	71280	58200	41170
1 7/8	48320	36240	96640	83680	68320	48320
2	52760	39560	105520	91380	74600	52760

* Always use poly lines in these short lengths, with a swivel or parallel (double) lines

** Swivels should be of ballbearing, free acting design. Use heavier shackles for multiple lines

Use common sense on sizes; a 2000 lb. generator can quite safely use I" steel cable under an S61

Manufacturers' working loads are normally expressed as a percentage of new rope strength, not normally exceeding 20%. Note also that you cannot tell when a rope has been shockloaded unless it breaks (shockloads are sudden changes in tension, exceeding 10% over the working load). Synthetic fibres can "remember" the effects of shockloading.

For horizontal loads (cylinders, etc.), lines should be 1.5 times the distance between the attachment points on the load.

Wire Rope Diameter (Inches)	Single Vertical Hitch	Single Choke Hitch	Single Basket Hitch*	2 legs/ basket 60°	2 legs/ basket 45°	2 legs/ basket 30°
3/16	650	480	1,300	1,100	900	650
1/4	1,150	860	2,300	2,000	1,600	1,150
5/16	1,750	1,300	3,500	3,000	2,500	1,750
3/8	2,550	1,900	5,100	4,400	3,600	2,550
7/16	3,450	2,600	6,900	6,000	4,900	3,450
1/2	4.700	3,500	9,400	8,150	6,650	4,700
9/16	5.700	4,200	11,400	9,900	8,050	5,700
5/8	7,100	5,300	14,200	12,300	10,000	7,100
3/4	10,200	7,650	20,400	17,700	14,400	10,200
7/8	13,750	10,300	27,500	23,800	19,400	13,750
1	17,950	13,450	35,900	31,100	25,400	17,950
1 1/8	22,750	17,000	45, 500	39,400	32,200	22,750
1 1/4	28,200	21,200	56,400	48,800	39,900	28,200
1 3/8	34,800	26,100	69,600	60,300	49,200	34,800
1 1/2	41,300	31,000	82,600	71,500	58,400	41,300
1 5/8	48,600	36,400	97,200	84,200	68,700	48,600
1 3/4	55,900	41,900	111,800	96,800	79,000	55,900
1 7/8	65,400	49,000	130,800	113,300	92,500	65,400
2	72,600	54,500	145,200	125,700	102,700	72,600
2 1/4	90,300	67,600	180,600	156,400	127,700	90,300
2 1/2	111,800	83,700	223,600	193,600	158,100	111,800
2 3/4	131,100	98,200	262,200	227,000	185,400	131,100

For 2 Choker Hitches multiply by 3/4

For Double Basket Hitch multiply by 2

*with vertical legs - a basket hitch is similar to a strop

Table values above are for slings with eyes and thimbles in both ends, Flemish Spliced Eyes and mechanical sleeves. For hand-tucked spliced eyes, reduce loads according to the table. For eyes formed by cable clips, reduce loads by 20%.

NETS & STROPS

Nets are used for loads consisting of many small pieces, and are also very useful for killing lift with those that can act like a bucket as you fly (like an Argo or a bunch of canoes). On a 206, a net about 10-12 feet square with a 2" square mesh is quite suitable. Items should be carefully and evenly stacked in the centre, with the net stretched round the load on the ground before pickup. Individual

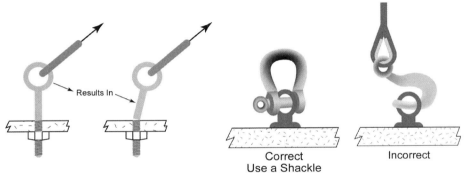

Results In

Correct
Use a Shackle

Incorrect

Never insert the point of a hook in an eye bolt:

light loads, such as jerrycans or containers, ought to be lashed together, since the net may not completely enclose them at the top. If there are many small items, consider a tarpaulin as a liner, which will stop them falling out, but don't consider it much! It could come loose and wrap itself round a sensitive part of the helicopter.

Strops are otherwise known as web slings, made out of synthetic material such as nylon. They are not used as primary slings, but tend to be hooked together at the top with a shackle.

Loading and Unloading Areas

Non-involved people should be absent, and there should be no loose articles to be blown around by the downwash and cause damage (it is possible for slash to snag and inadvertently operate the manual release). Approach and departure lanes should be into wind.

Performance

Some helicopter flight manuals allow a higher maximum gross weight for external operations

Max gross weight performance is determined by the HOGE graph

You should consider the site to be in hostile terrain if it is difficult to make a successful engine-off landing or protect the occupants from the elements, or if there is minimal SAR capability for the anticipated exposure. Under those conditions, a twin-engined helicopter might be better (depends on the country. In some places the above conditions are normal!) In calculating an OEI hover capability, forecast winds should be ignored under 10 kts, and you should only allow half of any above that anyway.

Once you have calculated HOGE performance for your density altitude, whether single- or multi-engined, the load should then be reduced by a further 10%.

 The HOGE capability allows no margin for manoeuvring in the hover and the vertical drag of the load, so the maximum normal operating weight should give HOGE with a 10% Margin. If precise positioning is involved, or there is poor visual reference, or turbulence, try 15% (you will be doing a towering-type takeoff to stop the load dragging, so you need power in hand). A ceiling of 600 feet and 3 nm visibility should be regarded as the minimum weather conditions.

ORIENTATION OF EYE BOLTS:

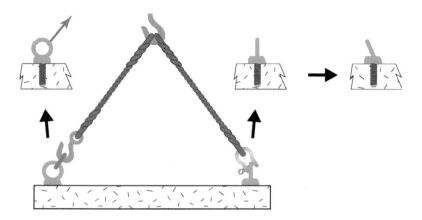

Correct Orientation is
in the plane of the eye

Incorrect Orientation
When the load is applied to
the eye in this direction it will bend

Your fuel consumption will be higher than normal, because you are using high power in the hover and flying slower (leaving the heavier loads till last will help with your planning).The C of G will always move closer to neutral with a sling load, and farther from it when the load is released, so ensure that the C of G limits will not be exceeded either way.

If the pressure on the surface is less than standard, you are effectively at a higher altitude, and your machine will not fly so well. To calculate the likely pressure altitude of a location beforehand so you know what your performance will be, get the local altimeter setting, find the difference between it and 29.92 (or 1013), convert it to feet (1"=1,000 or 1 mb=27 feet at sea level), then apply it the *opposite* side of 29.92. For example, for a helipad on the side of a mountain at 400 feet above sea level, with an altimeter setting of 29.72, your PA at that location would actually be 600 feet, since the difference between 29.92 and 29.72 is .2, or 200 feet *added*, and where you would enter your performance charts, since they are calibrated for standard atmosphere (the altimeter setting is *below* the standard pressure, so your answer should be *above*). Just put the big numbers to the bottom. You could also get PA from the altimeter itself, by placing 29.92 or 1013 in the setting window, and reading the figures directly.

The V_{NE} also varies with the situation and density altitude. For example, according to the flight manual, you can fly a sling load under the Bell 206 at 91 mph. However, remove the door and the V_{NE} is now 80 mph. Thus, although some loads might be able to fly at high speeds, other restrictions may stop you (I never plan on getting more than 60 knots from any load). Two sheets of plywood and two ten-gallon drums in a net may fly at 30 or 40 mph before they start swinging uncontrollably, so you must fly as slow as necessary to keep the load under control (also, reduce V_{NE} drastically with an empty lanyard on the hook, so it doesn't get tangled up in the tail rotor). From experience, if you're lifting an argo (amphibious thingy), which spills air out like a bucket, you won't get much more than 40 kts anyway.

Preparation of Loads

The weight of each load should be known, with sand and stuff kept dry and, if possible, weighed immediately before loading, as a good soaking will increase the weight dramatically and give you a surprise when you lift it. What happens most often, however, is that the guys just bundle stuff into a net, and as long as you don't overtorque the machine you're assumed to be OK (of course, you might not be!)

There should also be no possibility of the load, or part of it, becoming detached in flight. Loads that depend on the integrity of nailed parts, weak metal or wooden components must be avoided, or include a secondary method of support. Heavy dense loads attached to a wooden pallet may well become detached from the pallet in flight, even when carried in a net, although an exception is pallets of bagged material, the weight of which is spread over the whole surface of the pallet and wrapped to it. Delicate parts must be protected against damage and impact when lifting/lowering.

Unless the flights are very short and over low risk terrain, shackles, etc. should be secured with locking nuts or wire to stop them loosening under vibration.

Slings should be positively secured to the load, and not rely solely on weight or friction to maintain their position or security. They should also not be subjected to sharp edges or excessive angles which may cause their SWL to be reduced (see *Effect Of Sling Angle*, above).

TYPES OF LOAD

There are 4 types - *rectangular, cylindrical, heavy compact* and *nets*, and five ways of lifting them, starting with nets and ending with a four-point sling (through 1-, 2- and 3-point). The ratio of height, width and length, and the weight of a rectangular load will determine if it will rotate, fly broadside to the direction of flight or oscillate. Cylindrical loads generally fly quite well if they have a reasonable ratio of size to weight, but bulky, light cylinders can oscillate quite badly and may need a very low airspeed to keep them manageable. Logs (or long thin cylinders) are most stable when slung vertically, but will also fly well horizontally from a two point sling. Heavy Compact loads may travel at speeds up to V_{NE} without becoming unstable. They are often the most comfortable loads to fly.

Note: Lifting points already attached to anything may have been designed for cranes, which don't, as a rule, fly sideways or get caught in updraughts, etc.

Although convenient, and mostly used for nets, a **single-point hookup** is not always the best plan, particularly for short lines, and should only be used on loads designed for it, with a swivel, to stop the line unravelling (such as a pole or log). You also need to be particularly careful about the load's C of G, or it may tip and start rocking, so keep the heavy end down.

For a single pole or log, wrap the rope or chain *twice* around the end and carry it vertically (steel rope grips best). Short objects should *not* be lifted by their centre on a single line, as they can tip vertically if the balance is wrong, either due to the line not being centred or from aerodynamic forces.

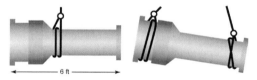

Use a two point hook-up instead, which is a common method for long loads, like drill collars, pipe stems, lumber, etc. In fact, it is probably the most common hook-up on the oil patch for any long load (here is a double choke hitch):

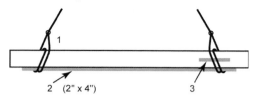

Double wrapped cables an equal distance from the ends (see above) with an over-the-load angle of 60° have little tendency to slide unless the pick up is very poor, especially with some wood between them and the load to stop them slipping (3, above). If the lanyards are of different lengths (longer at the rear), the centre of mass lies behind the suspension point, for more stability, and the load flies nose-high (if it's nose-low, the air hits the top of the load and makes it oscillate from side to side - using lanyards of equal length results in a nose-low attitude in flight because the wind resistance drags the load behind the machine to produce a tilt to the line).

Three-point hookups are not common, and are usually used for loads designed for them, but boats are lifted this way. **4-point slings**, on the other hand, are quite common and used for box-like loads, attached to each corner. Where the sling may catch or damage the load, use spreader bars to keep the rope away, as well as for stability (left). However, beware of loads with a high C of G when slinging from pick-up points on the base.

Loads rigged 4-point with an apex angle of more than 90° (used so you can remain in ground effect) may cause severe control problems from a harmonic resonance between the stretched cables and the rotor system.

A short cable between the hook and sling solves this.

Big flat loads of plywood or timber, or with large top surface areas (huts) are a pain to fly, since they can easily create their own lift, and get artificially heavier from the helicopter's downwash. Also, the middle layers tend to slip out, so make sure there's something on the ends to stop it (the same will happen when you pick up multiple logs).

Tag lines (short lines underneath a load) are for loaders to grab on the setdown. They should have a safety latch and be heavy enough to stop waving about, but long enough to catch if the load is swinging too fast.

Argos and Quads can be moved with strops underneath. Just wrap a strop around the Quad engine (two is better), ensuring it goes inside the pedals (or they will get bent). Then choke it and attach it to the hook.

Personnel Briefing

All concerned should be aware of:

- The hook-up
- The setdown
- Hand signals
- Proper use of radios (i.e. don't hog the airwaves)
- Direction to move in case of engine failure*
- Not standing under the load
- Number of trips between refuelling stops
- Retrieving slings and nets
- Use of protective equipment
- Accident procedure

*This depends on which way the rotors spin. With North American rotation (anticlockwise as viewed from above), the helicopter has a natural tendency to drift to the left when the engine stops, so ground crews should always work to the right of the helicopter and be prepared to go that way if they think anything untoward is happening. Astars go to the right. I've yet to find the person who can detect an engine failure from the outside before he gets squished, hence the preference for longlines.

HOOKING UP

Ground crew are nice to have, but you often have to do this yourself. If you have one, the marshaller should be at least 25m from the load with his back to the wind so you can see him from your high position. If he needs to change places, he should cease marshalling first, so he doesn't move backwards into unseen obstructions. Using standard marshalling signals, you will be positioned over the load, where the loaders apply the static discharge probe to the hook and place the eye of the net or sling inside it. As mentioned, loaders should always work to one side of the helicopter, and should also keep clear of the exit and approach paths, in case you have to drop a load.

If you don't have a long line, you will have to get the belly hook within the loader's reach, which is one reason why the mirrors are there. Don't get so low so that he has to crawl around on his knees, because that restricts his manoeuvrability and takes up more time, and you might snag the load with the aircraft. Ensure the lanyard does not go over a skid and that the hook-up man is clear before starting to gently tighten the lanyard.

The loaders then give an affirmative signal to the marshaller, who gives you the "move upwards" signal until the slack has been taken up. You will increase hover height slowly, until the strain is taken, with the loaders guiding the strops as necessary, taking care to be free to move away quickly should the need arise. At all times if the engine fails, the ground staff must move in the appropriate direction. They should not turn their back on the load, or get directly underneath it, or the flight path. Neither should they wrap lines directly around their wrists or bodies. When finished, they should clear the area as soon as possible.

As you take up the weight and the rope stretches, the difference in performance will immediately become obvious - it will feel as if you're attached to a large rubber band. Once you're hovering, and the marshaller is sure that the load is clear of the ground (and you are sure you can lift safely, flashing the landing lamp once to indicate this), the marshaller should check behind you for other aircraft and give the affirmative signal, as you will find it difficult to check for yourself. Keep a close eye on your Ts and Ps at this point - if you don't have a power reserve, the load is too heavy and you will have less control at the destination, especially if it is higher.

Tip: A load should "spring" slightly into the air, or at least come off cleanly (experience will tell you the difference). If it just about makes it, or is a strain even to get it to move, don't do it.

Once off, the machine will feel quite sluggish, as if it's tied to the ground. Move upwards slightly in a towering takeoff, and nudge forward slowly, giving due regard to the load's inertia, without alternately slowing down and speeding up, or you will confuse it. Rather, move forward and keep going to allow the load to follow, which sometimes takes a bit of courage, to see how it flies. Make all control movements *smoothly* and *evenly*, keeping the downwash inside the rotor disk. Try not to allow the load to sink, as, if it hits the ground or gets tied up in a tree it will trip you (there is a natural tendency to sink as you go forward anyway as you gain translational lift, hence the towering-type takeoff). The torque used at this point will give you a good idea of what is needed for landing, so be careful if you are going to a higher altitude (for example, it reduces at about 2-3% per 1000' in a 206).

Once in flight, remember that the load is the part that should be kept straight and level, not the helicopter, and keep away from anything underneath that could be damaged (well, try anyway!) Where possible, keep low for a few minutes to be certain the load will fly well as, if problems occur, there is less altitude to get rid of to set the load down. *Only with imminent danger to the aircraft should a load be jettisoned*, and this is usually from excessive swinging (commercially, dropping loads is regarded as a non-macho thing to do, but it's your backside the helicopter is strapped to). As a point of interest, 10 gallons of fuel from 500 feet will go straight to the basement of a 3-storey house. If you drop anything obnoxious in water, expect your company to pay for the clean-up and testing - your responsibility lies well beyond the confines of the helicopter; actually as far as a load can reach if you drop it.

DO NOT play with the buttons on the cyclic!

LOAD BEHAVIOUR

External loads increase the frontal area of the aircraft, which naturally increases (parasite) drag, so you will need more power overall, which will reduce whatever margin you allowed in the first place, and your manoeuvrability. A load may be easy to lift, but present enough drag to cause severe difficulties, particularly where you reach power limits too quickly to maintain forward flight, and the load overtakes you and pulls you along (doubling the parasite drag means you need three times the power). A longline needs more anticipation, so you need a high degree of coordination and patience. It's not the sort of thing that can be learnt in any other way than with lots of practice.

Every load has its own V_{NE}, unfortunately usually only found by experiment, which is why you should always start off slowly and build up to a point where it starts to give trouble, then back off by about 5 kts, as low as possible, so there is less height to get rid of in a hurry. Most helicopters will carry loads at quite high speeds, but the load itself might not be able to handle it - a sudden input of drag when something falls off could become quite a problem, and even concrete blocks produce lift. Although customers don't like to pay for unnecessary flying, there's no rush. Take it easy. Also, remember your machine's V_{NE} with the doors off!

 The V_{NE} for an external load set by the manufacturer is usually derived from the aircraft's response after pickling the load. Mostly, there will be a strong pitch up when the load is released because the collective will too high. The max speed is seldom based on structural parameters, as most are low relative to the normal V_{NE}.

Unevenly shaped loads will tend to spin and, if they're slung without reference to their centre of gravity, could tip over. A drogue chute can stabilise them, but use a windsock type rather than a pure parachute, which will bounce around trying to spill the air out (you could always punch holes in it). Naturally, these must be kept well away from the tail rotor. Logs or cut timber (and canoes!) usually fly poorly unless a tail is installed, which can be made out of a bough or piece of plywood, so it sticks out of the back.

Scaffolding and planking can swing violently with only a few knots' change in airspeed, and aerofoil shapes could even generate their own flying characteristics. Bulky loads with a tendency to float, like empty containers, will benefit from leaving doors or panels open, to reduce drag and keep the load facing in one direction.

If you have to join two lines together, they will fly better empty if you have the shorter one on the belly hook. They will fly even better if you can stop the join from moving or flexing with a couple of splints and some duct tape.

Vibration

Oscillation or excessive vibration can come from a number of places, usually a combination of the stability characteristics of the load and forward speed. Heavy or dense loads, such as bags of cement or drums of kerosene, will not usually present problems, but large-volume loads of low density can oscillate at a certain critical speed, again, usually only found by trial and error. You can dampen any oscillation by reducing your airspeed to at least 10% below this critical one, going slower if necessary while increasing power. Turning could provide enough centrifugal force to stop it as well, which is also the usual remedy when the load starts to swing, but this will increase its effective weight, possibly to more than your lifting capabilities (a good reason for not being too tight on payload), so applying centrifugal force in these cases could make things worse.

Swinging

In general, the longer your cable, the less will be the chance of any serious oscillation. For a longer cable, if the load moves for an equal distance, the angle involved will be smaller. As well, any disturbance will tend to get damped out a lot more before they affect the hook.

If the load starts swinging in straight and level flight, you are likely just going too fast. However, whatever the cause, you should be able to stop it inside two moves.

The load will want to stay under the hook, due to the influence of gravity. If you move three feet to the right, for example, the load will pause while its inertia is overcome, accelerating it the same way until it passes under the hook and continues to about three feet the other side, to form a pendulum action (it is always a good practice not to rush the takeoff and to lift without a swing in the first place).

You could wait until the load is almost at the end of its travel to the other side of the machine, and then move over it, but then the load won't be where you want it to be (although it might be where *it* wants to be!) A better tactic would be to wait till it almost gets back to where it started from and move the helicopter towards it, laterally, and cross-controlling if necessary. Pulling back on the load just makes things worse.

Thus, load swing is proportional to speed and the length of the sling - the faster you go, the more it occurs because of the load's own lift and drag. Put simply, the load will always move further in a swing than the helicopter does - if you move 4 feet to the right, the load will travel 8, after a short pause, as it moves 4 feet to the other side of you. As it takes as long for the top end to swing through 15° as it does for the bottom end to go through 45°, like any pendulum, if you move the opposite way, as is natural, you just make the load swing faster. Going with the swing, that is, load to the left, helicopter to the left, will stop it quickest (do it on your side so you can see what's going on, but it works both ways). This moves the fulcrum point from under the helicopter to over the load, which removes the directional vector. If you want it to go in a particular direction, wait till it is going that way before correcting it, then you will be almost pushing it into place. Put more simply, don't pull the load the way you don't want it to go, and try not to use the cyclic!

You can do the same with a fore and aft swing, but watch out! Very often, a load can suddenly produce more drag, if it turns, for instance, which makes it slow down and pull on the helicopter to put the nose down, which you try to correct with aft cyclic that simply puts the tail rotor nearer the line. If the load starts forward again, and pulls the nose further up for you, correcting with forward cyclic may set you nicely up for mast bumping, or at the very least leave you hanging with the stick right back. The correct thing to do here is to apply collective, which will also add an upward force, assuming the clouds aren't too low, but using a longer line will help, too.

Tail rotor pedals can also be used to stop a lateral swing, especially as their misuse may have started the swing in the first place! One method is to apply some pedal in the direction of the swing as the load comes in underneath the helicopter. Another (from the RAF Puma guys) is, when the ball is out to one side, use the opposite pedal *when it starts to come back* (not before). Lowering the collective is often a good tactic as well, assuming you have enough height and are at low speeds.

SETTING DOWN

Approach into wind as much as possible, coming into the hover high enough not to drag the load. You're best to undershoot rather than overshoot, as it's easier to creep up to a target than go round again if you miss it, although going around is often a better tactic with a load on the end of a longline. To preclude an airflow change making the load unstable, slow down before descending.

Because of inertia, manoeuvres should be anticipated well in advance and made smoothly (not suddenly) with reference to the speed of the load over the ground. In a confined area, the load will tend to pull you down as the wind effect is lost, so a couple of knots in hand under

these circumstances may be desirable. Keep a constant scan going, because you need all the information you can get, especially when it comes to depth perception. Having the doors off, especially in an AStar, really helps with peripheral vision, even if you can't get your head out of the door. If everything goes pear-shaped, it's because your scan has stopped, as it might if you get fixated and suddenly tense up on the controls. Relax and start looking around again, it will soon get better.

As with any sling load, keep the disk as flat as possible, and particularly practice the transition from forward flight to the hover. The more coordination there is between the controls (keeping the nose absolutely in one place), the less incentive the load has to start swinging - if you haul in the collective first, then adjust the pedals, it will go off like a rocket.

Once in the hover, you again come under the guidance of a marshaller, who signals descent until the load touches the ground and the cables become slack (if you haven't got a marshaller, you can judge your height from whether the ground crew are looking up or down, or use the ground shadow, if there is one). Release the cables *after moving to one side* so they do not foul the load or hit someone on the head. Don't drop cables from anything more than normal hover height, and especially not under too much tension, or you'll get somebody in the eye (also, whatever is in the hook shoots downward quickly, and the hook itself will be opened violently and may be damaged). A manual release is provided if the electrical one doesn't operate, and, once it has done so, you should see a "load released" signal from the marshaller (see left), whereupon you hover by the side of the load while the replacement sling is placed inside the cabin, having moved away vertically first. Behave at all times as if the load has not been released.

If you can't hover, keep max power in and let the aircraft settle, without overtorquing - you will only be pulled down as far as it takes for the load to reach the ground, so just try and give it a gentle arrival. If you are delivering the load by yourself, land behind it, as far back as possible so you don't get the sling under the skid, and to provide a little tension for the hook mechanism to work.

There is a phenomenon called *Collective Bounce* that occurs when a sudden vertical force is placed on the helicopter, making you think the load has reached the ground. Although really relevant to larger machines, it can occur on smaller ones, and arises when the resonance of the blades matches that of a vibrating rope. The collective movements to correct this get out of phase, due to the response lag, and the answer is simply to stop moving the collective or go into autorotation, as the machine will self-destruct about the fifth bounce. A little extra control friction will help.

VERTICAL REFERENCE (LONGLINING)

Longlining is slinging with a line over 75 feet long, with a remote hook at the end. Although it's not the complete answer*, many pilots prefer longlining, if only because problems with the load occur further away from the aircraft, and produce less hassle with the controls and tail rotor (and downwash doesn't artificially increase the load's weight or throw up dust. In fact, in the desert, you need a line at least 100 feet long to keep you out of the muck that will otherwise clog up your engine and get inside the inner workings of the main and tail rotors). You're also that bit further away from mechanical turbulence, although almost always out of ground effect and right in the avoid curve, which may cause a legal problem. One big plus is that, if your engine fails, the ground crew have more of a chance to get out of the way. Another is that the delivery point doesn't necessarily have to take the helicopter as well** (of course, the reason why pilots like slinging in the first place is that there are no passengers!) According to some, all you have to do is put yourself into vortex ring a few times with a short line on, and you will be sold on a longer line!

*With a crowning fire (with only a small area in the smoke at the head that you can get your helicopter into), longlining can be dangerous, because all you will see is a bucket flying around - you certainly won't see the line, and the pilot won't see you, being above the smoke. As a result, some authorities have banned its use, at least in concert with short-lining, or horizontal reference.

**Be aware that, in the final stages of inserting loads into tall trees, everything goes dark! (Make sure the trees are not taller than your line is long).

Using a long line will allow you to work into wind more often, but if you are working a very steep slope the line will need to be very long!

It's very challenging, but rewarding, flying, especially when you can place the hook in the hand of a guy on the ground who is 100' down in the trees and not even looking up for

it, in one fluid motion (on a nice clear day in Canada, when the trees are changing colour......)

To longline successfully, you have to maybe look downwards out of the door or a bubble window, and not through the front with a mirror, which is the more traditional method, hence the use of the term *Vertical Reference* (having said that, in some machines, notably the Astar, it's quite difficult to look out of the door, and the weather often means you need the doors on, so using mirrors is sometimes the only way, unless you have a hole in the floor - it's the practice in Europe). The problem with mirrors is that, although the load moves in the correct sense fore and aft when you look at it in the mirror, laterally, it works the opposite way round.

The long line has a hook at the end, which is inside a metal brush cage, to both protect it and provide weight when you are flying around with only the line attached - extended hooks without cages are dangerously unstable.

Note: Although you can use a long line with any helicopter, true "longlining" is done by yourself with your head out of the cockpit. The times when you might want to do that are with wide loads when you can't see it all in a mirror anyway, and in areas where you need a wide field of view. Plus, of course, with heavy rain on the mirror!

Note: Don't do any long-lining near high voltage lines or thunderstorms.

The window mentioned above looks like this:

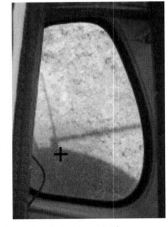

The cross represents some sort of datum point, which will change with the type of load and length of the line. It is a very valuable tool.

To establish it, hover over the coiled line and increase the hover height until the hook is just off the ground. Once any swinging stops, note the position of the hook in the window, as above. It is valid for the current line and wind conditions, and will certainly be different for a short line.

When picking up, keep the datum position over the load and you won't find yourself misplaced when you get to the point you are starting to lift (in contrast, with a short line, the aircraft will centre itself over the load). On setting down, place the Datum Point over the drop point and the load potentially about 10 feet above it when it catches up. Lowering the collective at that point will cancel out any residual swing.

Anyhow, as the extended hook needs an electrical supply, there will be an extra cable to control it, taped to the longline or incorporated directly into it. It will have a couple of spare feet at either end, which should be kept well away from the manual release.

 The electrical cable, even though it is only attached to the line with duct tape, can still pull your helicopter out of the sky!

At the aircraft end, a male and female electrical socket will be used to connect the longline electrically to the airframe, and will be held together with insulating tape.

You may also find a couple of changes to the airframe, in the shape of bubble windows, or instruments being repeated to the side (don't kick them on your way in) so you can see them. Because of the concentration required, it is very easy to lose track of time and what your instruments are doing. It takes about two seconds for the eyes to refocus from a load 150 feet away onto the instruments only two feet away, and the same time to refocus again on the load, so checking the instruments in their normal position will take your attention away from the load for about four seconds. Thus, instrument training that helps you read several at a glance could be useful.

Even then, you are likely to only cut the time down to about three seconds, which is even more of a reason to have a good power margin, or get used to the sound of gearboxes getting near to the overtorque stage (don't ask!)

A metal line is made of two strands of wire rope, wound around each other, to prevent spinning. Kevlar (or Spectra) ones are extremely light by comparison, but just as strong, although they can stretch. The only real pain with them is ground handling, because they have a tendency to get tangled up into knots if they don't have a protective covering (one suggestion is to get a large bucket, place the hook in and loop the line around the inside). However, they can also pick up twigs, etc., and grease from drills is apparently not good for them - they can also snag on trees. Being light, they can be hard to punch off in those circumstances, and tricky to position for the crews in a wind when empty.

You need to be particularly careful not to kink any line, which means not driving over it, landing on it, dragging it along the ground, or dropping it from great heights.

Depth Perception

Above about 40 feet (some say 15), it is very difficult to judge height properly, and handling depth perception needs some practice. It is for this reason that you should always look at the load through the same medium (preferably none at all) so you give yourself the best chance, because you can't judge height with monocular vision. For example, you are not helping yourself by looking at the load with one eye directly and the other through a door panel (or a helmet visor, or tears), which is one reason why the Astar is problematical. You could use flat sunglasses (polarised ones are good for depth perception, but make it hard to read the gauges) but they are easily blown off in the slipstream. Also, if the load gets hidden from one eye by a skid, that eye will stay focussed in line with the other and whatever is blocking its view will be out of focus and edited out by the brain, thus screwing up your depth perception. Because your eyes are focussed to a close object, you will perceive any movement of the helicopter as a much greater angular movement than if you were looking further away, which makes it difficult to avoid overcontrolling.

Having the Sun on the opposite side of the machine to you is a useful tactic, so you can use the load's shadow to tell its height from the ground better (shadows also help with your depth perception). Many pilots judge the height of the load above the ground by the time it takes their eyes to focus between the target and the load - the shorter the time gets, the nearer the load is to the ground. You should keep your eyes moving from the load to the ground anyway to keep your eyes active.

Getting Started

Aside from not having so many places to look out of, the Astar's door is further away from the front seats. As a result, it needs a minimum line of 100 feet (some say 130) just to see the load through a hole in the floor between the door and the seat, if you have one. However, even then you only really see it when lifting, or on delivery (below about 15 kts), which is why you need a mirror as well (I once had to do a job with a 75' line on a TwinStar, and had to deliberately swing the load so I could see it, at least once in a while! In this case, long tag lines helped the ground crew to catch it). Some machines won't allow you to look out with shoulder straps on, or when wearing a helmet, so try it all out on the ground first - particularly make sure, if you have the doors off, that you wear nothing baggy and you either roll or tie your sleeves up, or they will get in the way when they puff up and you won't see anything. Note where your hook is and see if you can make control movements with reference to its position. Try to wear a helmet, though, as even slow slipstream has been known to pull a headset off (and baseball hats, sunglasses, etc.......)

Naturally, there is some skill attached to longlining, and it isn't too hard to learn the basics, but it is something you need to keep current on, that is, *practice*, and be *smooth on the controls, moving them all at once*, to minimise the possibility of the load swinging. Also, watch for vortex ring when delivering the load, as you have minimum speed with power on and a high rate of descent if the load is pulling you in. Anticipation is the key, but you can only learn this after some experience, wherein lies the Catch-22 of needing experience but not being able to get it. Fire chasing offers the best free training, as you are often out on your own, and nobody is using a stopwatch. Longlining proficiency to some customers involves putting a small load at the end of a 100 ft line onto a 4 x 8 ft sheet of plywood three times from different approach angles, or even putting a load into a barrel, which is OK as far as it goes - the real catch is doing it quickly and efficiently, and this is something that takes time. Take it step by step and *do not* try to run before you can walk! This is one occasion where taking the long route when you're learning is definitely the quickest!

There are three variations on the longlining theme:

- *Operational longlining*, which can be done by just about any competent pilot and is fairly undemanding, provided there is plenty of time and a reasonable margin of power available, subject to a couple of caveats which are mentioned below - it's when you are operating to the limits of the load and the machine that the real expertise comes into play, but even this is nothing more than good downwash management and smoothness on the controls, coupled with anticipation. There is little accuracy or speed involved with operational longlining, and it can be regarded as just an extension to normal slinging - it's commonly used in fire support, where you dump a water bucket's contents into a relay tank, or pick up the hoses and equipment after the excitement's over.

- *Production longlining* is the fast and efficient movement of materials from one place to another, typically used in seismic work, where you try and drop 30-40 bags full of equipment an hour (logging and oil exploration need you to carry a few hundred loads per day, so the number of approaches is doubled, when you include the empty hook). This is fast work (but never fast enough!) often in places where people can't get around very easily, even without a 250 lb bag, and you will not be popular if you drop the stuff in the wrong place. Although GPS is useful here, many pilots (including myself) prefer to map read and get really familiar with the area before starting. In my opinion, the cockpit is not the best place for your head when buzzing around trees. In any case, the ground crews should mark the drop off spots with something like an orange X, at least 6 feet in size, with a double one at each end of the line. **Tip**: Ensure your own ground crew have the serial numbers of all equipment you move - this will stop the customer unloading any old junk on the insurance if you have to drop anything. **Another Tip**: Be very aware of vortex ring - production work involves floating in on minimal power and pulling it all in on very short final, for a high rate of descent with power applied. There is a fine line between speed and efficiency and recklessness. See the discussion below.

- *Precision longlining* usually involves moving drills, etc. because they are heavy and cannot be moved once they are on the ground. In this type of job, inches count (actually, an eighth up or down), as you are often holding equipment in place while people bolt it together. The usual precision for drills, etc. appears to be around 30 cm, although this depends on the guys on the ground - some like you to be precise, some don't bother

The real finesse with longlining comes when moving loads that take up nearly all the payload available, although you should never use it all, because you leave yourself with no margins, either with power or pedals. Some companies (and customers) expect you to "inadvertently" use more power than the maximum to get the load moving in the first place - that is, it's well known that you are not allowed to intentionally use more torque than that in the Limitations section of the Flight Manual, but you can do so by accident. What you do is up to you, but that margin is for getting you out of trouble, like when a load sucks you into a hole and you need to give it a gentle landing - you should always aim to do the complete job within 100%, which is what (HOGE) performance graphs are for, talking of which, remember that humidity can reduce their figures by as much as 10% or more, so be careful after a good shower. Put more simply, overtorquing (within limits) is for landing, not taking off, but you knew that already. Another consideration is looking after your engine - many turbine failures result from pulling too many cycles from minimum to maximum N_G, so if you don't need 100% torque, don't use it. It's also best not to reduce the collective lever to the bottom when descending, either, and to make power changes gently, avoiding over- and undershoots.

It's when an experienced longliner gets on the controls that the process becomes like poetry in motion, with the load and helicopter becoming a symbiotic pair, when every ounce of performance is extracted from the machine, even to extent of bouncing a load against a tree to set it in motion (without damaging it of course!)

Naturally, with the top half of your body twisted round, you have to learn some new motor skills. Some people say there is a tendency to pull the cyclic the same way as you are leaning, and back, but I found that a bigger factor was the drift that occurs when you lower the collective, which you learn to cope with automatically when learning to hover. To take a LongRanger as an example, from being nicely positioned over the load with a vertical line, and reducing power, unless you make a conscious correction, you will find yourself very much to the left of the load very quickly. It's too easy to accept the resulting parallax position as the normal one and try to take off again with a slanted line, which means a potential for dragging the load, although, in the final stages with a short line, the helicopter will position itself as the line gets taut. This is something you therefore need to practice, that is, keeping the hook over the load and the line in sight whilst pushing the controls in a strange direction. If you're likely to get the same training as me (none), get into a high hover, without a line first of all, and practice spot turns with your head out of the door, keeping the belly hook over the remote hook (it is *behind* you!)

Note: Much of this is contrary to basic flying techniques. You need a safety pilot!

Try a bit of ascending and descending as well, watching out for the left drift, but this is better done with a line on, which you can use as a reference. Remember, when you tilt the disk to move one way or another, that the lift vector is shortened, so you will need a little collective (another good reason for having a margin, as this will be hard with 100% torque applied!)

Next, get used to landing and taking off by looking at the back of your skid, so you avoid the transition from vertical to horizontal reference (if you do have to transition, do it as slowly as possible, from vertical to lateral to horizontal, and *vice versa*). Make any yaw corrections with reference to the skid (or wheel). Maybe do a circuit or two, as well. This also helps you if the engine stops, with no time to start looking horizontally - as mentioned above, just moving your focus from the load to the instruments, or the other way round, takes about two seconds each way, let alone reorganising your whole body and, speaking of engine failure, your chances of getting away with one are slimmer, as you will likely be in the avoid curve over inhospitable terrain, because that's where longlining is used. One technique is to dump the lever, hold the attitude, and pull up sharply when the pucker factor comes in, being prepared for forward movement. Cushion as normal.

Then practice approaches. Line a bug smash up with the aiming point, and fly every approach at the most accurate constant angle you can, with minimal power changes and pedal input (in a Bell 206, at least, 250 fpm is a good rate that will give you hardly any collective movement in the final stages).

Next, graduate to a line with a lightly loaded cargo net, followed by a heavier one, finally working up to an unloaded hook. You will find that, up to a point (where your power margins are eaten into), the heavier the load is, the easier longlining is. Get into a ten-foot hover, with the load over a point, then pick two points slightly apart and start landing the load on each one in rotation, in smooth, controlled movements (the less time you spend in the hover, the less the load will swing).

There is also a natural tendency to tighten up on the controls, the same as when you start mountain flying the first time. The same advice applies, however, which is to RELAX! If you have to take your aggression out on anything, do it on the collective, as the key to good longlining is proper downwash management, and spilling it with jerky cyclic movements does not help at all. This is one reason why there are holes in bearpaws, to let the downwash through. Even a ski basket can upset the airflow enough to spoil a lift. Remember the reason that control movements should be smooth on any helicopter - the induced flow takes a split second to build up, so if you do things too quickly, you get the drag without the lift.

So, be as gentle as possible at all times. Small, longer, controlled movements are always better than larger and shorter ones. The reason you "fly the load" is to stop you focussing on the helicopter and interpreting its larger angular movements the wrong way, although you shouldn't forget to watch where you're going. However, the big problem with concentrating on the load is that it is very easy to start it swinging. If you look after the basics, such as good heading and yaw control, rate of closure or acceleration, line position, and approach angles, the line and load should behave themselves. Also, get used to the sight picture of where the line/load combination should be in a normal still-air hover. This is from Paul Johnson:

> *"Take a CD marker pen and put a few straight lines across the skids and cross tubes, anywhere you think will be in your peripheral vision, when looking straight down at the hook. Maybe mark each line with a number, say 1-10 along the skid and A-J along the cross tube…*
>
> *Go into an OGE hover (or have another pilot do it from the other seat) and look down at the remote hook. In this perfect hover, the hook might hang at the imaginary intersection of lines D cross-tube and 5-skid. Hold the hover, and try to imprint this sight picture in your mind, with the aid of the index marks. Remove the ones you don't need on the next landing and in future, anytime you're told to "fly the hook, not the helicopter", fly the sight picture, not the helicopter!"*

Having done the usual preparations, such as ensuring that the line isn't tangled, and all the electrics work, you might also want to take a note of the altimeter readings of the lifting and delivery points - just add the length of the line to the elevation, so you get an idea of when it's going to get taut. This means that, when learning, you only need to stick your head out of the window just before you lift, to make sure the line is straight and the load isn't going to hit anything on its way out. As mentioned above, performance charts are important, as a difference of 5°C

can make the difference between getting a load off the ground (or not) when operating to the limits. Although customers like to use the maximum payload, I still like a safety margin of somewhere between 10-15%, because pulling full power for long periods is not good engine handling (you need to watch your pedals, too). Remember that you have to get the load moving onwards and upwards (especially the latter), which eats into the maximum continuous power limits, so you really need to get going inside the first minute.

I find it best to get over the area looking out through the windows, and only look down for the final positioning moves, for which you have to get used to your line's position relative to the ground (use the altimeter - I do all the power changes, etc., by the time the load is around 20 feet off the ground, then look out). This also means you can see the instruments (i.e. torque) a lot better, and that you don't hit anything. You can look down more as you get proficient. I have also found it useful, for positioning, to extend my view forwards and sideways to prominent points and line myself up with them, which is useful when I can't see the landing point.

Another reason for the margin is to ensure you have enough fuel to get to the destination - running short of gas is one source of pressure you *don't* need. It's all very well for the sales department to tell the customer that your machine can lift a particular load, but you also have to get it somewhere! A stage length of 25 miles or so at 40 knots means over 30 minutes' flying, or more if the load flies badly. Unless you have fuel there as well, you also have to include the journey to the fuel drums, so minimum fuel should be avoided, and neither should you have to remove survival kits, etc. to make a job happen.

There is a certain springiness to a load as it comes off the ground which tells you it's a good lift. In my experience, if you have to struggle to get the load off, and it's reluctant to do so anyway, that's the time to think again. Reduce the payload, wait for the wind or a cooler day, or whatever, but STOP. With an AStar, things are a bit more black and white, in that it will either pick up the load, or it won't. A Bell gives you a little scope for coaxing here and there - as mentioned, beeping down on a 204/5 by a couple of RPM can help it lift more, but that also makes the tail rotor work harder. Usually, when the machine stops on its way up and it looks as if that is as far as it's going to go, wait a minute, and she will produce a little bit extra.

Once the load is airborne, immediately ease the nose forward, after a little collective, with the aim of proceeding smoothly upwards and forwards. If it is out of the door, your head should be constantly moving between the forward and vertical positions, to make sure you don't lose situational awareness, which is another way of saying don't hit anything! Once you are clear of obstacles, you can start thinking of turning, and can put your head back inside the office and proceed as for a normal load (that's the time to spit out all the bugs that have splattered all over your teeth). It's also important to ensure that you don't overpitch, and that you remain in balance as much as possible - if the machine wants to weathercock, let it. Don't forget, *reduce power only when you have the speed and height you want.*

Getting It Back On The Ground

Well before the landing site, start slowing down. The mountain technique of using the collective to do this works well, and is also worth some practice (small left and right adjustments are also better made with the collective rather than cyclic.) A slight crab will not only help to keep the target in sight, but the drag from the more sideways presentation of the fuselage will also reduce the speed. Your scan at this point will be off the scale, especially if you are dropping off to a point higher than the ground (that is, on a platform) and you don't have the shadow to give you an idea of where the bottom of the load is. The approach should be slightly steeper than normal, with extra caution to help your depth perception and to avoid dragging the load into obstacles or allowing it to touch down too early. The steeper approach means it is even more important to be into wind and to keep the rate of descent completely under control to avoid vortex ring, especially if the load is heavy and your power margin is low and you get pulled into the hole.

Try not to overshoot, as the load will have its own inertia and will continue forward on its own (this is what's meant by flying the load). Backing up to reposition with a longline is always a problem, even with power in hand, as it's easy to set the load swinging (especially in an AStar). Large cyclic movements can make the load move a large distance, which is not good in a tight area. In this case, it's better to go around and try again, as there should always be movement, even if it's vertical (the exception being when you need to keep a casualty as still as possible so you don't smack him against the side of a cliff!)

Get the power in early and make your control movements slow and steady - one slow constant pull of the collective to establish hover power *just before* establishing the hover attitude prevents overtorques, large power swings, and large pedal inputs. Also, if something isn't right, you will know early in the approach while you still have some

height. If you flare and wait until you push the nose forward to get into the hover to pull in the collective, the aircraft will rotate around its C of G and start accelerating downwards, so you will need even more power to stop the descent on top of compensating for loss of translational lift. If you are pushed for power, a level transition to the hover uses less power than going forward and down. Less power also means you need less tail rotor, so there's less chance of LTE. In other words, start trading airspeed for sink rate (reducing both) at a very slow, but consistent rate, minimizing pedal input if possible. The idea is to try to fly the load exactly where it needs to go on the first approach.

So, you should at this point be coming in nicely on a long, steady approach, aiming for a point just above the elevation of the ground plus the length of the line. Keep moving forward and down, following the load. With power in hand, you could probably come to a complete stop, but still aim for a no-hover deposit, so you don't give it a chance to swing. Use all the controls at once, especially in an AStar - keep that nose straight. In the last stages, having got the load moving by itself, so to speak, follow it in, and get on top of it (that is, vertically overhead) to land it, in effect "pushing" the load into place with the line. You do this with the collective - lowering it starts the load moving out in front, and raising it lets you catch up.

Without much power, you have to aim for the target directly and place it in one movement, which only really comes with practice (about a week when you're starting will set you up nicely). You should be concentrating on the load's groundspeed, which should be around a walking pace. A hook should be delivered to the loader so that he can catch it, that is, at a low walking speed and within his reach vertically and horizontally (there is a great amount of job satisfaction in getting it just so that all he has to do is hook the load on). If he has trouble with the hook-up, you will have to hold the hook steady so he can play with it, without pulling to either side. After the hook-up is complete, give him time to clear and *watch him do so*. Then, and only then, should you gently take up the tension on the load.

In mountains, you have to learn to ignore the slope, and be aware that your downwash may well bounce off it and push the load away from where it should be, that is, underneath you, which is why you might sometimes use very long lines, when the slope is very steep and you still find your skids in the trees.

When the load is finally on the ground, do not just get rid of it. Many heads and many thousands of dollars' worth of equipment have been damaged by thoughtless pilots who released loads and allowed the rings or eyes to drop onto the unprotected heads or cargo.

After a job, when landing, curl the line up on itself, then pull back when quite close to the ground, ensuring it is away from your skids.

 Unhook the line when you shutdown, as it is very easy to take off and forget it is there.

Tips & Tricks

From various experts:

- Long lining, like any other area of flight, requires the pilot to be smooth above all. Many people get away with erratic control movements in other areas, but the consequences show up much more dramatically when on the line - especially in a Medium.... You want to be a "good" line pilot, then be smooth. Lots of people can fly with a line on, but the good ones are smooth and controlled.

- Make every pick up and put down as controlled as you can, put that hook on a rock, stump, drum, tuft of grass, anything - just make sure you have a goal. Even after a 10 hr day moving drills, I still make myself put that hook somewhere specific when I get back to my fuel pad. It's just like regular takeoff's and landings, why be sloppy on easy ones and only pay attention on toe-ins?

- Practicing is good, practicing well is better

- Move closer to the pedals (crank them back to you in a 206 or slide the seat forward 2 clicks in a 350).

- Fly in trim. When it starts to fall apart I concentrate on the pedals (everyone knows that as you change power you have to change the pedal position but it is easy to forget.)

- First you get good then you get fast. Watch the really good drivers - they are slow around the people on the ground. Once the load is really clear then you can honk on it. Nothing upsets the ground crew more than having you launch with their hand entangled in the net.

And one from Doug Potts about drill moves:

You never wrote about moving drills for drillers!

I am glad to say I haven't done this type of work in years and will avoid it at all cost for the rest of my days.

You see, drillers work in the drillagram world, unlike the rest of us! Your helicopter knows all about drillagrams even if you don't. A drillagram weighs twice as much as a gram, hence the name, but you should have NO problem with this as the last pilot didn't have a problem moving 2000 lb loads with a 206B!

And as you said in your book, the customer who is always right and knows more about flying helicopters than you will ever know just said you will have NO problems if you just get those dam drill parts moved from A to B and stop whining. Turbines are suppose to whine, not the pilot!

On one occasion I had gone into a Manitoba Hydro camp at Conawappa on the Nelson river to do a drill move. Oh, it was only a short hop so you won't need more than 20 lbs of fuel in your 206B, I was told by the drill foreman.

As you said, he has lots of flying under his belt - he is the customer. When I told him I would not even start the machine with 20lbs of fuel let along move sling loads for half an hour he knew he didn't like me. So I said "let me worry about how much fuel I need in the machine but how much does the tower weigh?"

His reply: it was designed as a helicopter flyaway tower so you should have No Problem.

I can see we are going to have a problem so I say OK, I will hook up to the drill tower with a 50 foot line and will be in contact on their radio frequency and let them know if it will go or not. So I put a little more fuel on board and up we go. They have one of their monkeys climb up to the top of the tower to connect the 50 foot line that I am using. I get over the tower and get connected. Slowly, I pull power and watch the torque gauge climb until it is almost to 100%. Then the foreman says: "OK, the bolts are out, take it away!"

I am tickling 100% already and he says take her away! So I say put the bolts back in as it is way too heavy to fly and you better hurry as I have only 4 minutes left to be pulling in the yellow on my gauge.

I am watching my watch and the torque gauge and the tower below and the fuel gauge which has already reached a state that I am not comfortable with. Then I tell our great foreman that if they cannot put the bolts in I will lay the tower down.

He gets very upset with this proposal. They finally get the bolts in and I land beside the sideshow. I have 15 lbs of fuel showing on the dial. and I am pissed. They are looking like puppydogs that made a mess on the kitchen floor. I said unless you can break down the tower it is going nowhere. And his reply?

"That's what the last pilot said as well!" Go figure!!!

TYPICAL HOOK LOADS AT MAUW HOGE

Assuming 1.2 hours' fuel, pilot at 185 lbs and slings, etc at 100 lbs.

2000'

º C	-15	-10	-5	0	5	10	15	20	25
AS350B2	2097	2097	2097	2097	2097	2097	2097	2097	2097
AS350B1	1846	1846	1846	1846	1846	1846	1844	1734	1624
AS350BA	1652	1652	1652	1652	1652	1652	1652	1539	1424
AS350B	1400	1400	1400	1400	1400	1400	1400	1400	1300
500D	1134	1119	1104	1089	1074	1054	1034	1004	974
Bell 206B	1002	987	972	957	942	927	912	897	882

3000'

º C	-15	-10	-5	0	5	10	15	20	25
AS350B2	2097	2097	2097	2097	2097	2097	2097	1990	1885
AS350B1	1846	1846	1846	1846	1846	1746	1654	1544	1434
AS350BA	1652	1652	1652	1652	1652	1589	1474	1359	1244
AS350B	1400	1400	1400	1400	1400	1400	1375	1250	1125
500D	1099	1069	1039	1009	979	949	919	889	854
Bell 206B	977	962	947	932	917	902	887	872	857

4000'

º C	-15	-10	-5	0	5	10	15	20	25
AS350B2	2097	2097	2097	2097	2097	2000	1895	1790	1685
AS350B1	1846	1846	1846	1794	1684	1574	1464	1354	1244
AS350BA	1652	1652	1652	1652	1544	1424	1304	1184	1064
AS350B	1400	1400	1400	1400	1400	1400	1250	1100	950
500D	1064	1024	984	944	904	964	824	779	734
Bell 206B	952	937	922	907	892	877	862	832	737

5000'

º C	-15	-10	-5	0	5	10	15	20	25
AS350B2	2097	2097	2097	2080	1945	1830	1715	1600	1485
AS350B1	1846	1824	1714	1604	1494	1384	1274	1164	1054
AS350BA	1652	1652	1634	1509	1384	1259	1134	1009	884
AS350B	1400	1400	1400	1400	1275	1150	1025	900	775
500D	1024	974	924	874	824	774	724	669	614
Bell 206B	927	912	897	882	857	827	757	687	617

6000'

º C	-15	-10	-5	0	5	10	15	20	25
AS350B2	2097	2085	1975	1860	1745	1530	1515	1400	1285
AS350B1	1744	1634	1524	1414	1304	1194	1084	974	864
AS350BA	1652	1652	1484	1354	1224	1094	954	834	704
AS350B	1400	1400	1400	1275	1140	1005	870	735	600
500D	984	924	864	804	744	684	624	569	494
Bell 206B	902	882	862	842	822	742	662	582	497

MOUNTAIN FLYING

5

In the mountains, general principles common to other areas will be vastly different. You must be prepared to adapt your flying techniques as the need arises, both for the peculiarities of the region and the type of aircraft. In other words, have not only Plan A, but Plan B, C, etc. up your sleeve because, very often, once you've looked at a site and gone round for finals, you will find that a cloud has got there before you! You cannot afford to assume that a particular situation is the same as, or similar to, any other you might have encountered previously, which is one reason for not leaving passengers for too long, because you might not be able to pick them up again. Sometimes, you can even create your own clouds by pulling down warm air from an inversion above. You can also expect fog, especially in the early morning, which will often stick to the sides of valleys for quite some time.

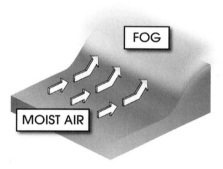

In the UK, mountainous areas include Scotland and Wales, the Lake and Peak Districts, and generally any hilly country above 1500 feet amsl, although a geologist would probably expect to see 2000. In many other parts of the world, these would be considered as just foothills, as one description of mountain flying includes a reference to 5000 feet density altitude.

In the USA, the boundaries of mountainous areas are defined in the AIM. In Canada, look out for *Designated Mountain Areas*, which include the Rockies, extending into the USA. However, air behaves the same way round pointy bits of ground whatever its height above sea level. The difference with real mountains is that you have less power to play with, hence the mention of density altitude (the other points to watch are wind and optical illusions). However, density altitudes over 5,000 feet can also be found in non-mountainous areas, like the desert, as can turbulence and mountain waves.

A couple of thoughts for when you're very high up; how long it takes to get down if you have a problem, and meeting anyone else at that height who doesn't expect you. And oxygen, or the lack of, leading to hypoxia. Also, throttle correlation does not work as well because larger collective movements are needed for relatively small butterfly displacement (the Robinson version disconnects at 5,000 feet because the collective is too high).

Super-cooled water droplets in warm moisture-laden air blowing up a mountainside through the freezing level will freeze on contact with anything.

SAFETY

Experienced pilots develop rules that they will not deviate from under any circumstances. Some points to bear in mind, include, but are not limited to:

- Mountains take no sh*t from nobody

- Always be able to turn towards lower ground, which means approaching or crossing pretty much everything at 45°, from around half a mile away. This is called *dropoff*, meaning an escape route.

- Never go past the point from where you can make a 180° turn without power and without hitting anything, especially in box canyons. This means thinking ahead because, if you leave it too late, you will go past that point while you make a decision. Slowing down allows you make tighter turns.

- Make all turns away from rising ground.

- Be prepared to use flight controls for different functions, such as the collective for speed, etc.

- Take off as cleanly as possible to avoid getting snagged.

Otherwise, keep clear of marginal weather* and avoid strong winds. Allow much more time for twilight, as it will be darker inside mountains sooner.

*Flying near the clouds keeps you further away from the ground, but it also gives poor visibility and you can get into trouble without seeing it, especially when updraughts force you into the cloud. It's best to fly in the lower third of the distance between the ground and the cloudbase.

PERFORMANCE

Pressure Altitude

Pressure altitude is the height in the standard atmosphere that you may find a given pressure, usually 29.92" or 1013 hPa, but actually whatever you set on the altimeter - if you set 1013 on the subscale and the needles read 6,000 feet, the PA *for that setting* is 6,000 feet. So what is indicated is the height of the pressure selected. PA is a starting point for any calculations for performance, TAS, etc. Thus, if an altimeter is set to 1013, it is measuring Pressure Altitude with respect to Mean Sea Level. In ISA conditions, Pressure Altitude is the same as True Altitude.

If the sea level pressure is different from 1013, obstacle clearance heights and airfield elevations, etc. must be converted before using them. To do this, get the local altimeter setting, find the difference between it and 29.92 (or 1013), convert it to feet (1"=1,000 or 1 hPa=27 feet at sea level), then apply it the *opposite* side of 29.92. You could also get PA from the altimeter, by placing 29.92 or 1013 in the setting window, and reading the figures directly. The significance of this concerns performance - if the pressure on the surface is less than standard, you are effectively at a higher altitude, and your machine will not fly so well. You often need to calculate the pressure altitude of a location so you know what performance you have available before you get there (power *available* for a turbine depends on pressure altitude. Temperature and power *required* depend on density altitude, below). As an example, for a helipad on the side of a mountain at 400 feet above sea level, with an altimeter setting of 29.72, your PA at that location would actually be 600 feet, since the difference between 29.92 and 29.72 is 0.2, or 200 feet *added*, and where you would enter your performance charts, because they are set for the standard atmosphere (the altimeter setting is *below* the standard pressure, so your answer should be *above*). You are *adding* because the sea is *lower*, and the figures ought to be higher.

Tip: To find what the standard temperature should be at any level, line up the two tens on the inner and outer wheels of your flight computer and take a look in the altitude window against the height you require:

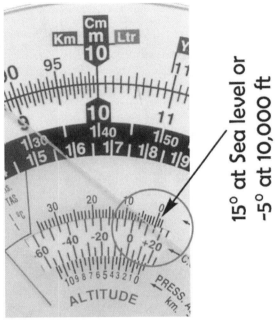

The rate of climb will be negatively affected by high temperature and pressure altitude, and *dirty rotor blades*.

To use performance charts effectively, you need to modify the pressure altitude you are really at for non-standard temperature by finding the ....

Density Altitude

As air expands (due to heating or being forced upwards), it becomes thinner. Thinner air is less dense (*Boyles Law*). On the surface, an increase in temperature decreases density and increase volume, if the pressure remains constant. At altitude, however, pressure reduces more than temperature does, and produces an apparent contradiction, where temperature will decrease from the expansion.

In other words., the air gets less dense when you climb, but it gets cooler as well, which offsets the effect a bit. Because of this, low-level operations (below about 5,000 feet) probably won't need you to get too concerned, apart from taking notice of airspeed placards and power limitations (some pilots report no real difference up to 8,500 feet). You should find that at least 75% power is available to a fair height.

As the main and tail rotors* turn at the same speed, as you increase altitude, higher pitch and power settings will be required and less of both will be available for reserves and manoeuvring (there is a higher chance of retreating blade stall). Sling loads, especially, increase parasite drag and restrict the turning envelope because they reduce the power available. In some helicopters, like the 500C, the rotor blades will stall before you reach the engine limits anyway. The dynamic pressure applied to the ASI is also reduced, so TAS will increase relative to IAS, and if you try to maintain a particular airspeed, you will be going faster than you think. V_{NE} decreases as well.

*Less pedal range will be available, so main rotor RPM *must* be maintained.

Larger control movements will therefore be needed, and there will be more lag, so the controls must be moved smoothly and gradually, or you may be on the ground well before that large handful of collective pitch even takes effect! Rotor RPM will rise very quickly with the least excuse, especially in an updraught, explained later.

For performance purposes, particularly with piston engines, if the air is thin before it goes into the engine, the combustion process becomes much less efficient, hence the use of superchargers (in a turbine, you can get near the temperature limits of the engine, which will run faster at altitude anyway). Air density reduces by 0.002 lbs per cubic foot (2½%) per 1000 feet in the lower atmosphere. Profile power reduces, because the air is less thick and offers less resistance, but induced power increases because of the increased angle of attack that is required.

The density altitude is the height in the standard atmosphere where you will find the density you are actually experiencing - 90° (F) at sea level is really 1900 feet as far as your machine is concerned. In extreme circumstances, you may have to restrict your flying to early morning or late afternoon. The idea is that the more the density of the air decreases, the higher your aircraft thinks it is, and the less efficiently your engine and rotors will perform. In the lift formula, you will see that the lift from an aerofoil is directly dependent on air density, as is drag.

$$L = C_L (\tfrac{1}{2}\rho V^2) S$$

The only things that usually vary are the air density and Coefficient of Lift (everything else is constant in a helicopter). When density does vary, the angle of attack (i.e. C_L) must also be varied to compensate and ensure that you obtain the same rotor thrust. In the formula, if the density reduces, the angle of attack must increase to

balance the equation, which means there is less surplus available. In the engine, the power margin is reduced.

The chart below compares Pressure Altitude against temperature:

°F/C	60/15.6	70/21.1	80/26.7
1000	1300	2000	2700
2000'	2350	3100	3800
3000'	3600	4300	5000
4000'	4650	5600	6300
5000'	6350	6900	7600
6000'	7400	8100	8800
7000'	8600	9300	1000
8000'	9700	10400	11100
9000'	11000	11600	12400
10000'	12250	13000	13600

It shows that, at 6,000 feet and 21°C, for example, you should be thinking in terms of 8,100 feet. On a 42°C day in Dubai, at sea level, the density altitude is 4100 feet, which can take 50 lbs off your max all up weight.

However, piston engines work on a fixed volume, and are affected by density altitude only. Turbine engines, being thermodynamic, depend on pressure altitude and air temperature **as separate items**. 5,000 ft DA can come from either 9,000 feet with -40°C or 2,000 feet and +40°C, but the power available in each case will be wildly different. To find DA on the flight computer, set the aerodrome elevation or Pressure Altitude against the temperature in the *airspeed* window.

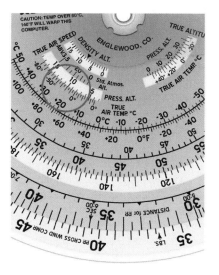

In the picture above, the temperature is -21°C at 10 100 feet. The indicated airspeed is 350 kts, and the TAS is 396. The Density Altitude is 8100 feet - quite a difference!

If you want a formula:

$$PA \pm (118.8 \times ISA \ Dev)$$

Multiplying the ISA Dev by 120 is usually good enough.

HUMIDITY

Adding water vapour to air reduces its density because the molecular weight of water vapour is lower (dry air is 29 - water vapour is 18). Engines are tested in normal air, and performance is not compensated for dry air, so "normal" charts include some humidity.

Note: 80% humidity means that the air has 80% of the moisture it can hold at that temperature, not that 80% of the air is water. The actual value may be only 1% water, depending on the temperature. Even at 100% humidity in the tropics, the percentage of water in the air is still relatively small. In fact, on cold days, humidity is less of a problem simply because cold air holds less vapour. A relative humidity of 90% at 70°F means twice as much than that at 50°F.

In addition, because around 80% of the air flowing through a turbine engine is used for cooling, the effect of humidity on a turbine is not tremendous, but it should still be taken into account, particularly in the wake of a heavy shower. As well, if you place a restriction over the inlet, such as a snow baffle, you will interfere with the cooling process and make the engine run hotter, which may affect your engine limitations at lower density altitudes.

Some rules of thumb:

- 0% humidity = baseline
- 25% humidity = +100 feet DA, and minus 1% power (therefore -1% gross weight to hover.)
- 50% humidity = 200 feet DA, minus 2% power
- 75% humidity = +300 feet DA and -3% power
- 100% humidity = +400 feet DA and -4% power

So, for each 25% humidity above 0, subtract 1% from your hover weight.

You can see that the change in DA due to humidity is low, at only 400 feet from between 0 to 100%, so the effect on the rotors is not much. In other words, the range of humidity between 0-100% makes a difference of only 400 feet in DA, so when you are not limited by the engine, that would be a difference of about 20-25 kg off your MAUW.

ILLUSIONS

There is a psychological aspect to mountain flying (you might hear lots of noises from the helicopter that you haven't heard before, the same as you do over long stretches of water!)

In the initial stages, it requires some self-control, as you overcome a certain amount of fear and tension, which is not good when you really need to be relaxed on the controls. You also have to cope with some optical illusions. Almost the first thing you will notice is the lack of a natural horizon, and will maybe want to use the mountain tops or sides as a substitute. This, however, will probably cause a climb, or other exaggerated attitudes, and make it difficult to estimate the height of distant ground, either from a cockpit or on the ground itself, so you will find it best to superimpose a horizon of your own below the peaks.

This is where using your instruments will help, both to keep your attitude steady and to give you a good idea of your height and speed (however, you're not supposed to be instrument flying!) An important illusion concerns landing sites - one that looks level with your current position may actually be higher or lower. The worst illusions are associated with cirques.

Below is a typical situation.

Superimposed Horizon

Notice the feeling of the ground rising in the near foreground which gives you the impression of increased speed, especially near to a ridge. Climbing along a long shallow slope is often coupled with an unconscious attempt to maintain height without increasing power so, unless you keep an eye on the ASI, you will be in danger of gradually reducing speed - if your airspeed is reducing, then the nose has been lifted or you're in a downdraught (downdraughts will be associated with a loss of height or airspeed for the same power).

Tip: Set straight and level power for whatever speed you want to fly at (e.g. V_Y) and pick a ground feature that is more or less where the horizon would be. Keep it in the same place on the windscreen and check the VSI - if you are climbing, pick another one further down - if you are descending, then pick one higher. This will be a great help with making level turns at the end of valleys and bowls.

AIR MOVEMENT

Up to a certain critical speed, flowing air will hug the shape of a body and be quite well-behaved in a *laminar flow*, after which it breaks up to form vortices. Airflow is laminar if it follows a smooth path and its parallel layers do not interfere with each other. That is, it is non-turbulent, and its layers have different velocities.

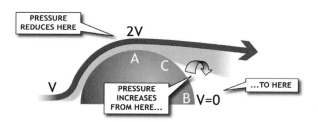

If the velocity of airflow as it meets a disturbance such as a mountain (or an aerofoil) is given a value of V, the velocity will be more as it goes over the top (at A). The movement of air over a crest line therefore has a venturi effect, giving an increased windspeed over the summit and a corresponding reduction of pressure, which could cause your altimeter to over-read.

However, the pressure remains higher at B, so the air cannot force its way there, due to the adverse pressure gradient. This reverses the flow around point C, where it is about to break away. Beyond C it forms into an eddying wake, which increases the resistance and turbulence.

The area of greatest lift occurs where the air is made to move sharply in a different direction.

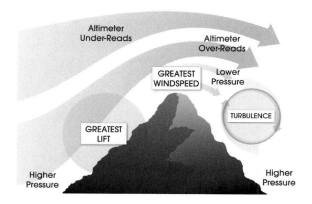

After passing over or round an obstacle, the air may become turbulent:

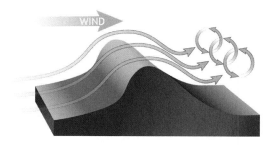

or form into rolls with a horizontal axis, over crest lines:

or gain a vertical axis, often associated with pinnacles or isolated hills where the wind will divide:

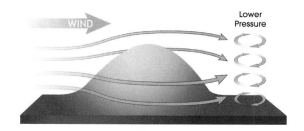

There will be a local decrease in pressure on the lee side, returning to normal some distance away from the range.

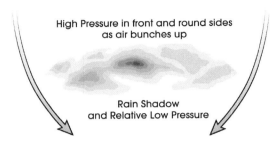

The venturi effect will also cause the wind to speed up as it passes through narrower channels or along valleys.

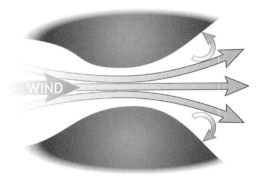

Thus, although the mean windspeed may be reported as 5 knots or so, you may find it as high as 30 in some places, and not necessarily coming from the expected direction.

A light wind blowing up a moderate slope with a regular surface will accelerate slightly to produce a gentle updraught. It will hug the line of the contour over the crest, then turn into a gentle downdraught. A roll may form near the bottom:

In a shallow valley with light winds, the air will follow the depression, producing gentle up- and downdraughts:

If the valley is steep, on the other hand, and the wind strong, the wind may blow across the valley and curl into it, producing a downdraught on the downwind side and an updraught on the upwind side - in short, the opposite to what you would expect:

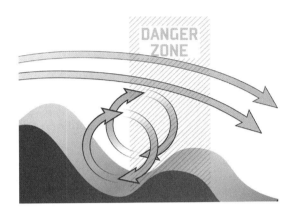

The Demarcation Line

Generally, upflowing air is smooth and downflowing air is turbulent, so you can often use turbulence as an indication of the vertical direction of the air you are flying in. The smooth flow will be broken if the ground gets rough, or there are trees, or the wind is strong. High winds and smooth steep terrain can be very laminar with a high downflow - on a glacier in warm temperatures, the boundary layer* will be thin, downflowing and strong (even 5 mph is 440 feet per minute).

*Airflow round mountains has a boundary layer, just the same as an aerofoil. The difference is that the air is moving, not the mountain. Thus, there is an area near the surface where the movement of air is relatively smooth and slower than the free stream, due to viscosity. It is important because we fly through it when landing or taking off.

The stronger the wind becomes, the larger becomes the area of downflowing air on the lee side. You have good approach possibilities from the front and sides of the hill before you hit the demarcation line, which is the point at which smooth air is separated from turbulent air where it breaks over a peak, like the transition point on an aerofoil. In the picture below, the blowing snow follows the demarcation line. Above or to the windward side (on the left), air is relatively smooth and upflowing - below, or to the right, in the lee, it will be downflowing and turbulent.

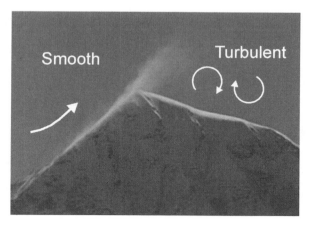

The demarcation line steepens as the wind velocity increases (and the severity of the slope), as does the area of downflow, which moves to the top of the hill.

It cannot get steeper than the terrain, and its base moves towards the wind as it gets stronger.

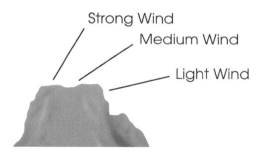

Tip: You don't have to keep the whole helicopter one side or the other of the line - many pilots keep only the blades on the windward side, as the fuselage by itself is not affected that much by turbulence. But this is a lot of work, and an easier method is to make sure you can see down the windward slope from downwind, and approach at 45° to get a safer angle.

In sunny conditions, the demarcation line can become less defined as anabatic winds flowing up the heated slope mix at the top of a peak. However, the sharper the rise in the ground, say with a cliff-face, the more it will remain near the edge and steepen with wind strength.

ROLLS & RIDGES

Ridges are crests that generally run parallel to the valleys next to them (see below). They may run in straight lines for many miles or be twisted into odd shapes. Ridges with narrow, jagged, sharp-sided, comb-shaped tops are *aretes*.

The general effect of a series of ridges is to form rolls between the crest lines, similar to the effect in valleys (above) if the ridges are deep enough.

As a result, on top of steep ridges there may be an area of nil or reverse winds which will be hard to find at first.

With sharp contours, turbulence will be more severe for the same windspeed, and they can produce areas of nil or reverse winds behind the demarcation line:

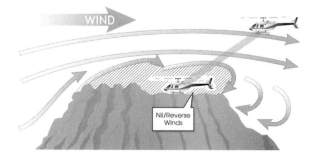

The safest area for landing will therefore be near the windward edge. You should only approach behind the demarcation line for a landing if you lose the wind and get forced onto the ground.

Even in light winds, an uneven slope will create turbulence on the upslope as well as the leeside:

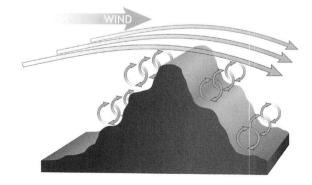

Winds

Winds can increase your operational ceiling, payload, rate of climb, range and cruise speed. They can also do the opposite, and be very difficult to predict, with formidable up- and downdraughts associated with them. When cruising downwind, along a lee slope or not, sudden wind reversals could make you exceed V_{NE} or even take away your airspeed.

There are several types of wind, which can be loosely be grouped into:

- **prevailing**

- **local**

    - valley

    - anabatic

    - katabatic

    - mountain waves, etc.

PREVAILING WINDS

Prevailing winds blow more from one direction than any other, and will be a characteristic of a particular region, so they are influential in the placement of runways.

As such, prevailing winds have a steady and fairly reliable flow, and can start to affect you from above the tree line, but more generally do so upwards from about 6,000 feet. Indeed, upper winds can come in many directions at different levels, and are usually the opposite of lower winds. Where mountains are concerned, they also acquire a vertical element, which is often where the boundary layer comes from. The weathering of surfaces and vegetation at altitude can give you a good clue as to the direction of the prevailing wind.

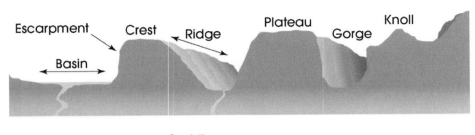

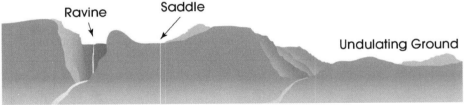

Note: A high pressure system may be strong enough to produce subsidence that will be stronger than the prevailing winds, so you may not find an updraught where you expect one! As well, subsiding air increases in temperature due to compression, and may form an inversion. This may be strong enough to prevent any air movement at all.

LOCAL WINDS

These are infinitely variable. Smoke from local fires can be used to detect their direction, as can water, but this may only give half the story - for instance, it's not uncommon for the windsocks at each end of Banff airstrip (in the Rockies) to be 180° apart! During the change of seasons, particularly between Winter and Spring, local weather factors can result in unusual winds at all altitudes.

Note: Using the movement of cloud shadows as a guide to wind direction is misleading because that at the base of the clouds will not be the same.

MOUNTAIN WIND
This is the prevailing wind interacting with a mountain range, or generally any wind associated with mountains.

VALLEY WINDS
These can arise from the prevailing wind flowing down a valley, or be generated by the Sun. They can be felt up to 2,500 feet above the valley floor, reaching their peak strength around mid-afternoon. A valley consists of three surfaces - the floor and two sides, and, depending on the time of day, and solar heating, each will have different activity. In other words, the interaction between upslope (anabatic) and downslope (katabatic) winds can produce a recognisable daily sequence. Before sunrise, the overnight katabatic winds produce a steady flow down the slopes and along the valley floor (it's often called a *Bergwind*). Soon after sunrise, the heated valley slopes* produce an upwards flow along the valley sides that are heated by the Sun until late afternoon, when it weakens and prepares the way for the katabatic flow in the evening.

*The East facing sides of valleys receive the energy from the Sun's rays first, and may cause the air on the West facing side to flow downwind as it tries to fill the gap.

KATABATIC
Cool air on a slope will flow down, because it is more dense, and therefore more subject to gravity, causing a *katabatic* wind. It's the same effect as you get in a closed room on a cold day, where there is a draught near a window even when nothing is open - the air next to the window is cooled, and flows downwards. The katabatic effect usually happens around sunset and overnight (when the heating effect of the Sun is lost), and its significance is not just that you might get some wind from somewhere you don't expect it (and downdraughts from severe slopes), but also that it slips underneath the air not in contact with the slope - if there is a river at the bottom of the valley, the extra moisture could also cause fog, so be careful when flying to valley sites in the evening. Katabatic winds tend to stay within 500 feet of the surface, and can arise quite suddenly. Glaciers have permanent katabatic winds.

Katabatic winds are often called mountain breezes. They can be stronger than valley breezes.

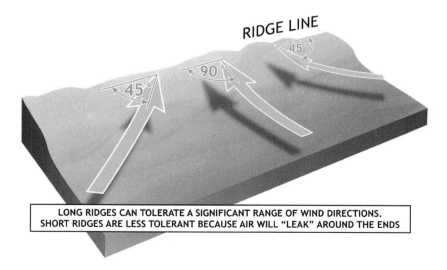

LONG RIDGES CAN TOLERATE A SIGNIFICANT RANGE OF WIND DIRECTIONS.
SHORT RIDGES ARE LESS TOLERANT BECAUSE AIR WILL "LEAK" AROUND THE ENDS

ANABATIC

An *anabatic* wind is a warm wind flowing *up* a hill, due to ground heating and air expansion during the day. It is not a regular thermal movement (with the whole layer moving vertically away from the slope), but rather a *slide* of the layer up the hill but, to get any lift benefit, you must fly close to the surface. For example, a 10° or 12°C temperature difference between a valley floor and the peaks of a range will be enough to produce upslope winds of around 15 knots in mid-afternoon.

OTHERS

Gusts are rapid changes of speed and direction that don't last long, whilst *squalls* do. A gale has a minimum wind speed of 34 kts, or is gusting at 43 kts or more.

A large body of water in a valley may develop the equivalent of **land & sea breezes**, which arise from a temperature difference between land and water. Air over land warms up and cools down faster than that over water, because land has a lower specific heat and needs less heat to warm it up. Thus, temperature changes over land will occur a lot more frequently.

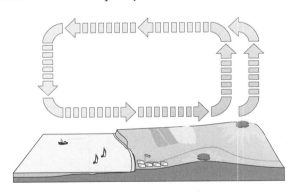

When the land is warmer, the air over it becomes less dense and the space left by the rising air is filled with an extra component coming from over the water (which is cooler, and does not rise) to produce a sea breeze which is added to any existing wind (in fact, a relatively high pressure is created at about 1000 feet over land, to produce a pressure gradient aloft).

With lower pressure at the same height over the water, there will be air movement towards it, at the upper levels (because the column of warm air is taller, and the relative pressure is higher), which will subside to come back towards the land. At night, the process is reversed to get a land breeze, but land is a poor conductor of heat and will only be affected through a shallow layer. Land breezes are weaker because the temperature differences are smaller and so is the local pressure gradient.

If a convergence is created, sea breezes can be strong enough to create their own cold fronts, well inland*, and even trigger thunderstorms, as the colder sea air undercuts the land air.

*In Australia, for example, sea breezes have been encountered 400 km away from the sea.

Knowing this is useful when you're going to a destination near water, and the wind (and landing direction) could be different than what you might expect.

WIND SPEEDS

As a guide to speed, whitecaps on water foam at 10 mph.

Turbulence

Aside from fatiguing the airframe, turbulence can cause loss of control, not only from forces being fed back through the controls, but blade stall can be caused by violent updraughts. It can arise from several sources:

- *Mechanical* turbulence comes from interaction with rough ground, and which therefore can be found on upslopes as well as downslopes.

- *Downflow* turbulence comes from wind breaking over the top of a hill.

- *Backlash* is a variation of the above which occurs when strong winds break over the top of a hill, especially if it has an abrupt (square-cornered) top.

- *Shear Zone* turbulence arises when different airflows meet, possibly between two valleys, but is more commonly frontal, to produce.....

- *Air Mass* turbulence, found when large bodies of air are forced to interact with high ground. The effects are felt well past the mountain range concerned.

Combat turbulence by maintaining maximum rotor RPM at climbing speed (V_Y). This will provide maximum power reserves, minimum pitch angles and least dissymmetry of lift, plus the least chance of retreating blade stall.

Downdraughts, etc

You can tell if you're in a downdraught by watching the airspeed (it reduces rapidly) and the position of the nose - if it yaws into the slope, the air is flowing down and *vice versa*. Induced flow is increased, reducing the angle of attack of the rotors, and therefore lift, assuming the collective isn't moved (and N American rotation), although some schools teach that it doesn't matter which way the blades go round. A lack of cloud above, i.e.

descending air, is also a possible indication. However, rotor drag increases, and the rotor RPM reduces, so applying collective to stop the rate of descent will only make matters worse. In fact, downdraughts can frequently exceed your climbing capabilities.

It's best not to fight them*, but go towards a lifting slope, or try for a cleaner column of air, maintaining climbing speed, plus half again, with maximum rotor RPM set.

*By diving in sink (above), rather than trying to maintain altitude, you are exposed to the effects for less time. Even though the rate of descent may be double or more the rate of climbing at V_Y, you will lose less altitude overall.

Don't expect help from the ground cushion - the effect will be less on a slope or grass anyway. If you are near the ground when you get caught in a downdraught, there will be the illusion of an increase in groundspeed. Don't be tempted to reduce it further!

Strong updraughts, on the other hand, can suspend you in mid-air with zero power - if the air subsides suddenly, you will be going down faster than you can apply it, which is one reason for using a shallow approach when the winds are strong, because you will always have power applied. With an updraught, the induced flow is opposed, and reduced, and the angle of attack increases, making you climb even further. Rotor RPM will also rise, because drag is less, and lowering the collective will make them rise even more. This will be aggravated by Coriolis effect from an increase in the coning angle.

Picture: Dark depressed puddles on water are called *bearpaws* (or catpaws) and are caused by downdraughts.

Mountain Waves

Mountain ranges are often too long for the air to go around them - it is mostly forced over the top to cause vertical disturbances.

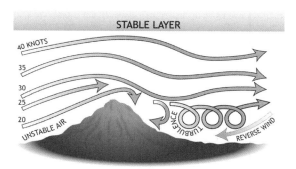

When such an obstruction has a stable layer of air above it at the 10,000 foot level (to act as a lid), and relatively unstable air flows over the range at about 20 knots, blowing broadside on (within about 30°), you can get standing waves for some miles downwind, because the wind has enough momentum to bounce off the ground behind the mountain and push the air already there out of the way. That air will fall again when it reaches a peak.

The vertical distance to which a mountain range can influence air movement is around 3 - 5 times its height.

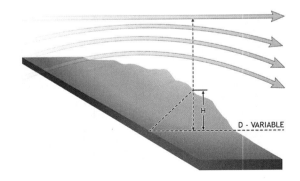

The horizontal distance, on the other hand, is variable, and cannot be estimated correctly.

Being standing waves, they do not move (although the air flowing through them does), and the distance between them is constant. They are easily identified by the types of cloud associated with them, which also do not move.

The wind speed and direction should be more or less constant right up to about 18,000 feet, although it doesn't have to be particularly fast over the peaks. However, it must increase with height. As the wind needs to be fairly straight in direction, warm sector winds and jetstreams can be very conducive to the formation of waves. They will be more dangerous in Winter simply because the wind speeds are stronger, and there will be a longer wavelength. There can be several miles between their peaks and troughs, which can extend between 10,000-20,000 feet above the range and up to 200 or 300 miles downwind:

The flowing air can be deceptively smooth, and only the VSI will tell you if you are going up or down. If you are flying parallel to a ridge on the downwind side in a smooth downdraught, as a result of the local drop in pressure associated with the wave, the VSI and the altimeter will not indicate a descent until you pass through a layer equal to the error caused by the mountain wave (they may indicate a climb for a short while), so you may not recognise that you are in a downdraught until you pass through the original pressure level which is closer to the ground than before you entered the wave. Thus, in cloud, or at night, you could be in some danger*. There could also be turbulence with accelerations up to 20 G in extreme cases.

*This does not just apply to light aircraft! 747s have lost complete engines in mountain wave downdraughts, but the most common problems are severe reductions in rates of climb and excessive rates of sink.

Downdraughts can be particularly dangerous into a headwind, as the airflow follows the general shape of the surface, and you will experience a strong downdraught just before the ridge. In other words, when into wind, height variations are out of phase with the waves.

They are usually in phase when you are flying downwind.

The potential loss of altitude is 500 feet if the wind is between 30-40 kts, 1000 feet between 40-50, 1500 feet between 50-60 kts and 2000 feet over 60 kts.

The trapped lee waves are associated with marked adverse pressure gradients as they go up and down, sometimes dropping over 5 hPa through just a few kilometres. There could also be large vertical increases of temperature (inversions) in the order of 10°C over 200 metres, and reverse winds, or strong increases in local wind speeds.

The combination of mountain waves* and non-standard temperature may result in your altimeter over-reading by as much as 3 000 feet!

An aircraft affected by mountain waves can expect severe turbulence below any rotors, downdraughts that may be stronger than the rate of climb and greater than normal icing in associated clouds.

*Technically, the wave over the mountain is a mountain wave, and the others are standing waves.

Watch for long-term variations in speed and pitch attitude in level cruise (the variations may be large). Near the ground in a mountain wave area, severe turbulence and windshear may be encountered, especially at the bottom of a rotor where you may get a performance decreasing shear if you are going in the same direction as the wind.

WINDSHEAR

This is the name for sudden airspeed changes over about 10 kts resulting from sudden horizontal or vertical changes in wind velocity - more severe examples will change not only airspeed, but vertical speed and aircraft attitude as well. Officially, it becomes dangerous when variations cause enough displacement from your flight

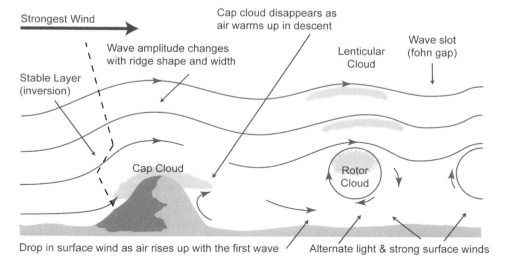

*The "stable conditions" are actually a layer of stable air sandwiched between less stable layers above and below

Strongest Wind

Cap cloud disappears as air warms up in descent

Wave amplitude changes with ridge shape and width

Stable Layer (inversion)

Lenticular Cloud

Wave slot (fohn gap)

Cap Cloud

Rotor Cloud

Drop in surface wind as air rises up with the first wave

Alternate light & strong surface winds

path for substantial corrective action to be taken; *severe* windshear causes airspeed changes greater than 15 kts, or vertical speed changes over 500 feet per minute. Expect it to occur mostly inside 1,000 feet AGL, where it is most critical, because you can't quickly build up airspeed. You can often tell the presence of windshear by clouds moving in different directions or plumes of smoke rising then going off at extreme angles.

Helicopters, especially, can suffer from windshear above and below tree top level in forest clearings, when a backlash effect can convert any headwind to tailwind.

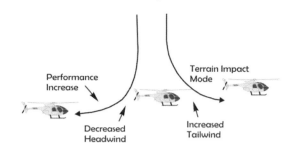

Performance Increase

Terrain Impact Mode

Decreased Headwind

Increased Tailwind

The most significant effect of windshear is, of course, loss of airspeed at a critical moment, where a wind reversal could result in none at all! You would typically get this with a downburst from a convective type cloud, where, initially, you get an increase in airspeed from the extra headwind, but if you don't anticipate the reverse to happen as you get to the other side, you will not be in a position to cope with the resulting loss. This has led to the windshear classifications of *performance increasing* or *performance decreasing (Microbursts)*.

The helicopter on the left in the picture above gets an increased headwind, so power is reduced to compensate. This takes effect just as the downburst is encountered, and the headwind becomes a tailwind, so IAS decreases.

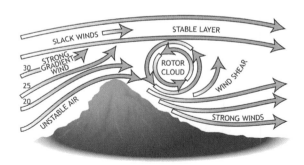

SLACK WINDS

STABLE LAYER

STRONG GRADIENT WIND

30

25

20

ROTOR CLOUD

WIND SHEAR

UNSTABLE AIR

STRONG WINDS

ROTORS

Rotors are ares of rotating turbulence found under the lenticular clouds that are a clue to the position of the peak of a wave. They are always in circular motion, constantly forming and dissipating as water vapour is added and taken away. They are dangerous, and the most turbulence will be found in them, or between them and the ground. Rotor clouds are formed in the same way as lenticular clouds, that is, from air forced upwards and condensing, then dissipating as they proceed downwards in the wave.

Rotor streaming is a phenomenon that can produce very severe turbulence, and violent up- and downdraughts together with considerable variations in wind direction and strength to the lee of a mountain range.

It occurs when air flowing across a mountain is enough to create waves, but decreases in effect above the mountain - that is, the wind speeds are only strong in the lower levels. The strong air flow begins to wave upwards at the range, but when it meets the slack winds, it shears back on itself to form a rotary circulation, assisted by the layer of stable air that acts like a lid on the upward flow.

The air downstream of the mountain still breaks up and becomes turbulent, but there are no lee waves, so the rotors travel downwind rather than stay in one place as they normally would. The rotary circulation causes a marked increase in wind strength downwind, violent up- and downdraughts, and severe turbulence. For example, in the lee of mountains in North Wales, a 25 - 30 knot gradient wind was turned into 50 - 60 knot wind. On one occasion, a Whirlwind with climbing power and speed set experienced a sustained vertical descent rate of 2 000 feet per minute in a downdraught and a 1 000 feet per minute rate of climb in an updraught, even though it was virtually in autorotation.

If the rotor forms in an inversion, warm air from above is rotated downward and heated further as it is compressed. On the other way up, cold air is expanding to cool further. Thus, very cold air ends up lying over warm air and conditions can become extremely unstable.

Watch for ragged cumuliform cloud if there is enough moisture present.

ASSOCIATED CLOUDS

As a clue to the existence of waves, you will see a *cap cloud* over the top of the range.

It is a good example of the Fohn effect at work, as the cloud disappears as the air on the lee side descends and warms adiabatically. A landing site on the upwind side of the ridge would not be usable.

A mother-of-pearl (or nacreous) cloud is a pancake-shaped cloud that is extremely thin and visible only for a short time after sunset or before sunrise when the sky is dark. It is normally seen in latitudes higher than 50°N, or over Antarctica. It is best seen in the polar regions at 80,000 to 100,000 feet when the sun is below the horizon.

The lenticular shown below may be found at the peak of each wave.

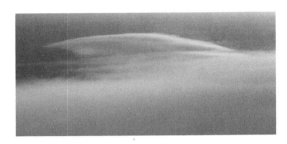

THE FOHN EFFECT

The **Fohn** occurs when saturated air is made to rise by mountains. On its way up it cools at 1.5°C per 1000 feet, and when it descends on the other side, having dropped its moisture, it warms at 3°C, so you get a dry, warm, downslope wind with clear skies. In California, it causes fires. Being downslope, it is a katabatic wind.

The essential point is that the temperature on the lee side of the mountain is warmer, which is why you can grow grapes in the Okanagan Valley in the Rockies.

Here are some sample figures:

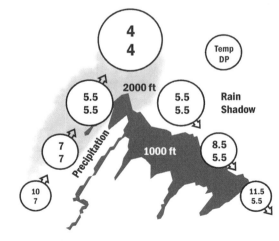

The cloudbase on the lee side is higher due to the precipitation on the windward side and the change in dew point. The temperature also increases through a greater depth than the cooling on the windward side.

The **Chinook** is a warm, dry, katabatic wind that comes off the Rockies, which are a huge physical barrier between the Pacific coast and the prairies on the Eastern side, from

Canada to the USA. Although fohn-like, the air on the windward side of the mountains is so cold (and therefore stable) that it does not rise up the slope, but is blocked. A trough of low pressure that is created on the Eastern side draws the wind at higher levels down the lee slopes. As this air has very little moisture to start with, it descends all the way down at the DALR, hence the warm and dry characteristics. Compression is also a factor.

VALLEY FLYING

When valley flying, upslopes or slopes exposed to the sun can produce updraughts, so place yourself on a converging course to the line of the ridge* and positioned to obtain a straight flight path two thirds up the slope and one across, which is generally the area of smoothest flight (it's always a good idea to hug one side of a valley anyway, to give you maximum radius if you need a to do a U-turn, and you avoid the shear areas in the middle).

*Some areas have traffic rules that require you to fly on one side of a valley. Even if that is the downflowing side, it will have a more consistent wind than the middle of the valley, which is likely where the novice fixed wing guys will be. Don't forget the radio calls if it is a VFR corridor (check out the diamonds below).

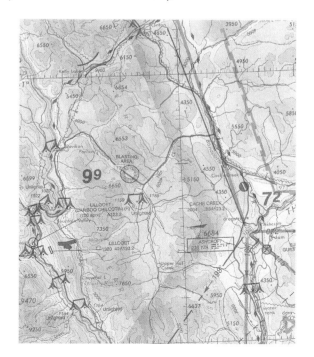

Two thirds up is about where the wind starts to think about going in the prevailing direction rather than staying local, and is where glider pilots find maximum lift:

Join the slope slightly high, at an angle of less than 45°, which should be reduced in stronger winds. Approachng from left to right allows you a better view of escape routes through your side window.

Tip: Do not become so mesmerised by the slope that you fly into any parts of it sticking out!

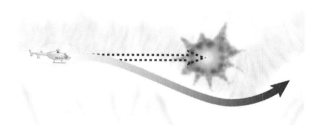

If you need to gain height in a particular area, it will be best to do so in something resembling a a figure of eight pattern*, rather than trying a spiral climb. This makes all turns away from rising ground (returning towards the site) to give you a good view all the time (going round in a circle means that much of the slope, and any landing points would be out of sight for much of the time).

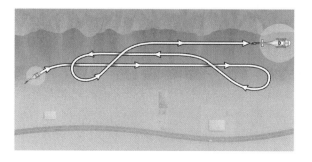

*The figure eight recce is the basic reconnaissance pattern for mountain flying. It is used when the shape of the ground near the landing site does not allow an adequate dropoff on all sides as is needed for the contour crawl described in *Finding The Wind*. That is, along the sides of long ridges, shoulders, saddles, and ledges, although the direction and strength of the wind may also be a factor. You can make all passes on the side where the landing site is, and take full advantage of any upflowing air.

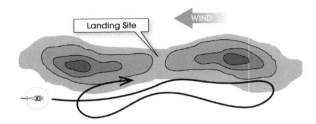

As ridges tend to be the top of two steep surfaces, they may have steep demarcation lines in moderate winds. The narrower they get, the fewer landing sites there will be. It's a good idea to fly up at a light weight first and check things out before attempting to land with a heavier load.

 When a ridge is approached from the upwind side (i.e. downwind) there is *usually* a cushion of air to help you up and over, if the wind is blowing more or less perpendicular to the ridge, but do not be tempted to use an updraught to take you up near the top of a ridge, to scrape over.

If you fly level at a hill with a 30° upslope, and have a tailwind of 25 knots, at 50 knots groundspeed your ASI will probably read zero as you approach the hill. The vertical component of the wind would be around 11 kts (1100 fpm) and the VSI would also read zero. So, you are not officially descending, but you will likely also be using less power than you would be in a hover. **Always keep the disk loaded** - never commit to a landing if your torque is very low on short final, especially in a 205/206 - the bottom will fall out on you as the fuel system is not fast enough to respond. The engines must speed up and slow down a lot as you work in turbulent air. Big blades make it hard for the governors to keep the rotor RPM at 100%, and your escape route is blocked by rising ground. If you are making a lot of large power changes, maybe think of going somewhere else.

There is also a downdraught on the other side, and potentially a lot of turbulence, plus a temptation to reduce airspeed as you get closer to the ground. In a 5-knot downdraught, descscending at 100 fpm, you may have to arrest a 600 fpm decent!

Passing over a crest line should always be done with a wide height margin - between 100-500 feet appears to be plenty. Again, approach at an angle less than 45°, but leave at a higher angle so you are out of the turbulence quicker.

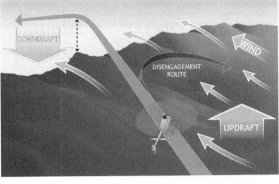

Tip: You can find out if you have enough altitude to cross a ridge by keeping an eye on the distance between two spots. The first is whatever you can see over the ridgeline and the other is pretty much anything else slightly beyond it. As you get closer, the spacing between them will decrease if you are not high enough. Put another way, if the terrain on the other side of the ridge is disappearing, you are not high enough.

Lee Slopes

You could also fly along a lee slope (that is, the other side from where the wind is coming from), taking advantage of the updraught formed by stronger wind returning on itself, called *riding the backlash**:

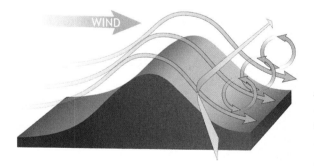

*Backlash has the wind curling back in the opposite direction than you would normally expect. The wind generally has to be over 10 knots to initiate backlash conditions and the event itself can be quite severe.

There is little room to manoeuvre, though, if something goes wrong, or you meet someone coming the other way. Smoothest flight will be obtained as possible to the ground, say about six inches, so you're in the boundary layer. This gives even less room for error, though!

Climbing over a lee slope must be started early so you can be high enough to avoid any downdraughts. You need an even greater height margin.

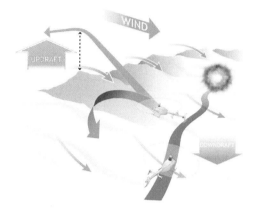

Remembering the discussion about the demarcation line, and the turbulence on the leeward side, a peak should be approached from either side of the turbulent area, which is handy, because approaches should be made at 45° anyway

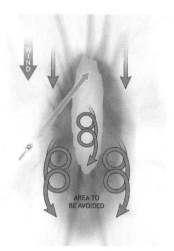

The Sun can make it difficult to assess angles and distances, particularly when going from sunshine to shade and back again. If the landing point is in shadow, try to make the approach in the shadow as well, and vice versa.

LANDING SITES

Wherever possible, landings should be made on ground that is higher than its immediate surroundings, so you can vary the approach according to the wind and have a clear overshoot and/or dropoff path. Customers, though, have an annoying habit of wanting to land on the most obscure sites, especially heliskiing guides! Try as much as possible to use the windward sides of a slope; leeward sides should be given a lesser priority, because wind flowing down the slope can increase its apparent angle (you need more lateral cyclic to hold the helicopter in place, and you could run out when you reduce power to lower the downwind skid). Also, you will not have the full effects of a ground cushion, if at all.

The wind coming over the peak will have increased in speed, due to Venturi effects, so a 15 knot wind can easily become double that, aside from your altimeter misreading.

In any case, the best visibility is down-sun, but it won't be all that bad with the sun at 90° to the aircraft. It will be worst when the sun is low on the horizon.

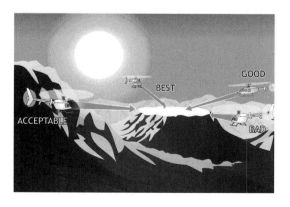

Log Pads and Platforms

Log pads are used when slopes are steep, or on rough ground. The quick and easy one is a single log across the slope for your rear skid to a solid mat of smaller ones:

They can be slippery! Platforms are still made from logs, but are much more refined. The problem with them all is that you can mostly only land one way, and there may be no room to turn once you get there, so approaching with the wind in the wrong direction is often the only choice. In such cases, you need much more anticipation than normal, and the willingness to throw things away much earlier. Of course, you don't actually have to land, but it's often worth a try. As with rigs or ships, it may be possible to approach to the hover nearby and move sideways on. Anyway, always be prepared to break off at any time, even if only seconds from success. Never commit yourself till the very last moment. Short cuts don't exist with mountains - they've been around a lot longer than you have! Mountain pilots always have a way out, even after they've landed!

Side hill landing pads can either be natural (a small ledge or shoulder), or man made (a plank or log pad). The natural variety must be checked out for a spot to land, allowing for main and tail rotor hazards. The man made variety leave no doubt as to where to land and which way to orient the aircraft.

It's best to approach a side hill pad along the slope to allow the maximum drop off and best overshoot. This also helps to avoid a high rate of closure and an abrupt flare. Man made pads on clear cuts often blend into the background and become invisible unless they are painted. Use a reference and note your altitude on your first pass. If the wind is moving up the slope you might run out of aft cyclic as you try to land facing the slope, due to the increased lift on the horizontal stabilizer and the aft part of the rotor disc. Even with no upslope wind, recirculation will cause the helicopter to drift toward the slope. This is

not a comfortable feeling - use caution!! Many man made pads are elevated and need OGE hover performance. If the pad hasn't been used in a while, ensure any brush or saplings haven't grown up to create a main or tail rotor hazard.

Note: Try to ensure that plank pads are sound and nailed down. They are not always level!

Log pads can only be landed on with the skids perpendicular to the logs. Most can only be landed on one way, usually with the nose pointed toward the hill or with the tail on the least hazardous side. Before landing, ensure the logs aren't rotten (be suspicious of a pad more than 2 years old), that the logs are secured, and the pad is in usable condition. If it is newly built or currently in use, such as with heli-logging, it must be checked for hazards every time you use it. If possible approach it at eye level.

APPROACHES

There are several schools of thought about approaches, but no real standard - as with many other activities involving helicopters, there is more than one "right" answer to this one. A fairly flat, disc-loaded (shallow) one will (in theory, anyway) give you the most control as you keep translational lift as long as possible, but there's very little power up your sleeve at the end because it is used to keep the helicopter flying, and you need to be very aware of your winds, as forward speed will mask the effects right to the last minute, although it does give you a good idea of the level of your site. This assumes that you remember your training and keep going forward and down, so the cyclic is ahead of the game and in the cleaner air in front of the machine that helps with translation. In other words, keep the rotor disc forward, so the flow of air is from front to back, especially where snow is concerned, but you shouldn't use the shallow approach with powdered snow anyway, because you could lose sight of your landing point at the critical moment in the resulting white cloud.

The other thing to bear in mind is that you are trying to land at probably the only spot available for miles around, and if you have a problem in a shallow approach, you aren't going to get there (with or without an engine) - in places like Papua New Guinea, when heavy at 12,000' or so, you are often committed from a mile away! However, a shallow approach does at least avoid a large flare and collective input in less dense air. It uses the least rate of descent and the power is used to support the aircraft rather than check its descent in the latter stages, so it is useful if there is a low cloudbase over your landing site or there is a severe updraught which pushes you up despite the collective being bottomed. However, if you get caught in a downdraught and are forced below your approach path, you may not have enough power or collective range to get back on it, and may have to go around.

You could, on the other hand, use a steeper angle, particularly if you're going into a clearing surrounded by tall trees, increasing the angle with the wind strength, but this may require large handfuls of power and attitude changes in the final stages if you don't get ground effect, so you wouldn't try this in an underpowered piston-engined machine that really shouldn't be there in the first place - the engine may be able to cope with it, but can the tail rotor? (leading with the pedals will help). Anyway, since ground effect reduces your (Bell 206) torque requirement for the hover by up to 15%, if you approach in such a way that you need no more than that amount to stop, you should find your descent stopping nicely in the right place, assuming the surface is conducive to it, and whether you have high skids or not. You also have some potential energy available for an escape.

I guess you could use whatever works - I generally turn in steepish (i.e. LS just forward of the nose), around 60 kts with the disc loaded as much as possible, consistent with descending at about 250 fpm, so power will be in early (saves a big pull at the end, and keeps you out of possible turbulent downdraughts on the lee of the hill). If the blades have some tension on them, they are less likely to be overstressed, and the controls are more responsive. At 250 fpm, the power used will give you a good idea of what you need in the hover, so you have an early chance to abort if you are using too much (you get to know with experience). This works, because 250 fpm reduces the thrust required to transition into the hover by about 15%, i.e. much the same as for ground effect and there should be minimal collective movement at the end. 250 fpm is about 20 feet every 5 seconds, if you haven't got a VSI.

Sometimes, you won't get anywhere near the top of a mountain without using updraughts. On strong wind days, the French (working in the Alps up to 12,000 ft) generally fly a climbing approach up the upwind side of the hill (having done a recce first from a safe height above the LS) and then almost spot turn into wind as they come over it. You can still fly back into the valley, on the updraughting side, if turbulence or power become a problem.

You could also approach at 90° to the wind, which will provide an escape route into wind and down the hill.

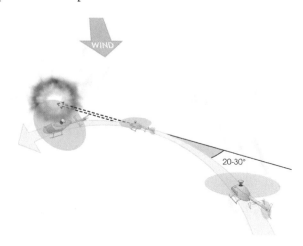

Whichever method you choose, if the machine wants to weathercock, let it - there's no point in using power or making a lot of effort to keep straight if you're going in the right direction anyway (and it helps keep the tail rotor clear). Keeping the whole of the windward side in view over the crest will keep you forward of the demarcation line and in the upflowing air. Coming in at 45° will help with escape routes and give you a better angle. When you make a final approach to land, remember that you may not be able to hover when you get there. If you do manage it, make it low, somewhere between 1-2 feet, and brief, one or two seconds. No-hover landings are not recommended.

As far as performance goes, if you can land at 6,000 ft in a Bell 47, you should be able to do so at 9,000 in a Bell 206. It's all in the wrist action!

Anyhow, whatever type of approach you use, do not approach any slope face-on - you will leave yourself no room for escapes or dropoff.

GENERAL RECONNAISSANCE

The basic procedure is carried out in two phases, the first being a general look involving an overshoot, and the second a more detailed one that may include the landing. If there is no wind, the procedure will be dictated by the ground - otherwise, the ruling factor will be wind and/or turbulence.

How high you do this depends on where the landing site is in relation to the surrounding terrain, the wind direction and strength, and whether there is any turbulence.

As height cues may be few and far between, you will need to use the altimeter to stop yourself descending below the level of the landing site.

The pattern used can be a variation of the two given below. This is the basic pattern for when you have reasonable power reserves and favourable conditions:

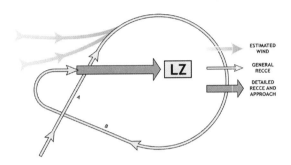

You can use the legs A and B to help estimate the wind strength and direction, and assess any up- and downdraughts, in addition to performing the power check.

The other pattern can be used if you have a difficult site, unfavourable conditions or marginal reserves of power:

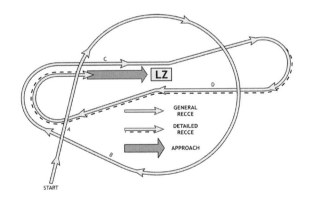

During either one, you need to note the size and shape of the landing point, escape routes, sun and shadow, etc., as discussed later.

POWER CHECK
See *Confined Areas*, later.

DETAILED RECONNAISSANCE
Here, you take a closer look at the landing site and further confirm the details noted before.

Finding The Wind

Finding the wind direction can be interesting if the site is bare and gives you no information, and it doesn't help that mountain flying tends to take place in high pressure conditions, that is, where the winds are light and variable (we are now talking about local winds, caused by convection, for instance, or katabatic effects, combined with the prevailing wind influenced by the ground, or even a mixture of them all). Even a cloud shadow can increase the speed of a downflowing wind from a cold surface.

One method of finding the wind is to slow down to V_Y, and fly **in trim** in one direction, then the other. The nose of the helicopter will point to any crosswind. Watching the groundspeed and noting the positivity of the controls will indicate whether you have a head- or a tailwind.

The **cloverleaf** procedure involves flying on a suitable heading and noting the drift, then turning 270° in the direction of the drift.

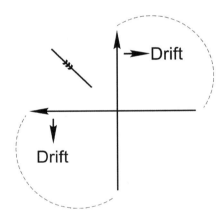

For example, if N was your first heading and you were drifting right, you would turn West. If you are now drifting left, the wind is from between N and W. If you were still drifting right, you would turn South and check again. Once you have narrowed the wind down to a 90° arc, split the difference and keep checking until you find a heading with no drift.

THE CONTOUR CRAWL (CIRCLE RECCE)

You can judge the effect of the wind on the helicopter, flying round the site with constant speed and power, or a constant altitude, which is otherwise known as a *contour crawl*, because you use one contour all the way round:

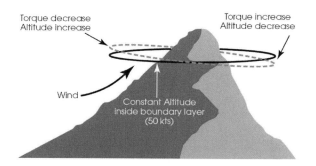

Torque decrease / Altitude increase

Torque increase / Altitude decrease

Wind

Constant Altitude inside boundary layer (50 kts)

It is a method of mechanically determining the wind information from multiple passes in different directions, comparing IAS to groundspeed, and heading to drift. It is usually used when wind information is not available from other sources.

After finding out what power setting gives you 50 kts straight and level, fly round the hill, at least 100 feet below the top (to keep in clear air as much as possible), looking at your power settings, whether the air is turbulent, your groundspeed varies*, whether you drift or whether the nose yaws into or out from the slope. How much pedal you use to keep straight is a good help - a lot of right pedal in a Bell means the wind is from the left, for example, and a fair amount of vibration through the pedals means it is behind you, but it may be a good idea, if you can't have it at the front, to get the wind off to the side that requires the use of the power pedal (the left, in a 206), in case tail rotor authority becomes a problem. Aft cyclic would indicate a tailwind as well. So, with constant power and airspeed (say 40-50 kts), when you rise, you will be on the windward side, and *vice versa*. On the other hand, you would use less power on the windward side if you kept a constant height.

*Maintain a constant IAS to check groundspeed properly.

 Use turbulence as a guide only in lighter winds - any found in updraughts could be from mechanical effects, like trees.

The Eye Level Pass

Having decided on wind direction in general, you now need to look more closely at your proposed landing site. In strong wind conditions, you won't need the contour crawl at all, because it's obvious where the main body of wind is coming from, but it may have very little influence over your final approach anyway.

As with any other potentially dodgy landing site, you need to check for *Size*, *Shape*, *Surroundings*, *Slope*, *Surface* and *Sun* (you don't want it in your eyes). The most important, however, in this case, is Slope, as there's no point trying at all if you can't land. You will get little idea of ground conditions if you overfly the site, so what you must do is have a look at eye-level, which results in the aptly named *Eye-Level Pass* (if the site isn't surrounded by trees).

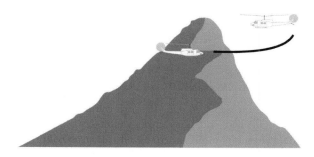

The most economical way is to start with a downwind pass, turn round and land, which is entirely possible if you know the wind direction before you start. Sometimes, though, this is not obvious at all, so just make an educated guess and fly at about V_Y in the direction you think is downwind very close to the site, *level with your eyes* (this point is crucial). As you do so, note the power used and the reading on the altimeter (those people used to using QFE may want to set it to zero), and climb up a little bit as you increase speed to about 60 knots, using the collective (you can maybe also do a power check). The climb is meant to stop you descending.

At 100 feet (some say 50), turn round *away from the hill* for another approach and repeat the process, taking note of the new groundspeed and deciding which way the wind is coming now you are closer. If there is no real difference in speed, check for vibration through the pedals, aft cyclic, etc, or anything that might indicate the wind is from behind. Any downflow will cause the machine to want to weathercock into the hill, and upflow will make it weathercock away.

Clues that you are downwind include:

- Moving fast, with a low tail

- Vibration, especially in the pedals

- Airspeed decaying rapidly, but still moving forward

- Fishtailing in the ground cushion

- You need more power

Tip: Be suspicious if you are using less than 50% torque - you may be in upflowing air.

The next step is an initial approach and overshoot, but if you have to make a circuit anyway, you may as well do another eye-level pass and get as much information as you can. Turn in at around 50-60 knots (at the 100 feet), taking particular note of escape routes, up and down draughts and turbulent areas. Maintain a constant angle, aiming directly for the point you wish to land on, controlling your speed with collective and avoiding any last-minute corrections. The idea is to keep the fuselage as level as possible, so don't move the cyclic at all, if you can help it. One reason for using the collective in this way is to minimise large control movements in the final stages, as this is a (very) shallow approach.

At 6000 ft if you lift the collective the RPM will droop as the engine is wide open so you must learn to make an approach with the cyclic.

Once inbound, if you have less torque for same stabilized airspeed you had on the initial downwind pass, you're in upflow/upwind and where you want to be. On short final, check the groundspeed and torque - both should be less. You have a loaded disk as you didn't need much of a descent. Then *fly* (not hover!) onto the LZ.

Maintaining A Sight Picture

Lining up a point beyond the landing site with the site itself can help you maintain a sight picture approach:

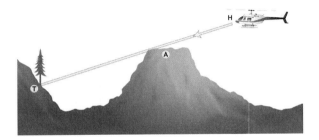

The approach is steady when the line between H, A and T is constant. You are getting too high in a situation like this:

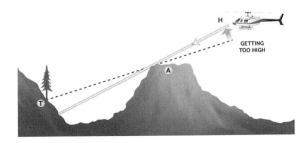

And too low like this:

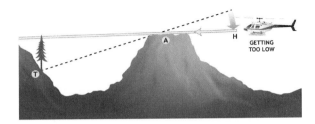

Being too low is the worst situation as the chances of turbulence and downdraughts are increased and you will need more power to get out of it (see *Shallow Approaches*).

Peaks

Pinnacles tend to be isolated, steep sided protrusions, although they are sometimes found along the headwalls of cirques. The wind has much more of an effect with pinnacles, in that you now have to contend with updraughts *and* downdraughts!

There is also generally no reference for closing speed on approach, which can give you the illusion that you are hardly moving. Then suddenly, you are arriving way too hot on short final. Use the double horizon technique and your altimeter to help your altitude control.

En-route you will be looking for the wind in that location and its effect on the demarcation line on the pinnacle. You will find that the demarcation line is very steep and will almost always result in backlash even if the winds are light (5-10 kts). If the winds are 15 kts or greater it may not be feasible to get onto the pinnacle safely at all – regardless of your gross weight! Also, keep in mind that the pinnacle can be affected by many different types of wind - Valley winds on the lower portion, thermal up flows on the sunny side (anabatic), down flow boundary wind on the shady side (katabatic), and prevailing winds at the summit.

This is one of the few mountain features where a circular recce can work well, but if it is an isolated pinnacle, you won't have any reference while flying away from it.

Even if the surface is flat, you probably won't get much ground effect here because the landing site will be small.

An approach to a peak can be either straight in or a descending turn into wind.

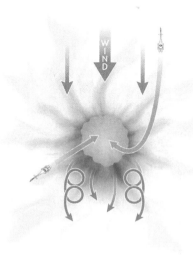

This is an example across the crest:

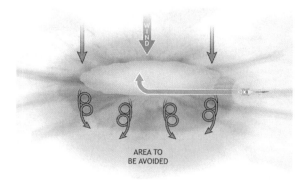

Cirques (Bowls)

Otherwise, known as a cwm (in Wales), or a corrie (in Scotland), a cirque is a bowl-shaped hollow at the head of a valley, formed from heavy glaciation that has receded, often leaving piles of rocks and gravel called moraines.

These can pile up to form a bench or series of benches near the headwall. Many cirques contain small circular lakes called tarns that have streams running out of them. These are formed when the ice melts.

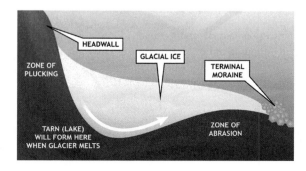

The rim of a cirque may be in the 10,000 foot range, with the bottom between 6,500 - 7,500 feet.

Note: Cirque landings should be avoided on windy days if possible*. The wind swirls inside them, and it is very unpredictable. If you must have a try, make sure you have a large power margin, an escape route, and do a few recces to see if the wind is the same each time.

*Once the winds are above 20 knots or so, every cirque, ridge, or mountain top will have a surprise! If the prevailing winds are blowing directly into it, the whole cirque will be in upflowing air. From any other direction, at least part of the cirque will be in downflowing air.

Otherwise, the upper prevailing winds descend into the bottom of the cirque basin and down and out of the feeder valley. Daytime heating can make air rise up the long valley floor and meet the downflowing air near the base of the headwall. This will cause rapid variations in wind direction, especially over a moraine or lower benches where the best landing areas are (they are elevated with a good drop off). With stronger winds, start with a recce around the rim, then half-way down, then by the proposed landing site.

Fly around the cirque walls in the boundary layer. If the situation is wild and unpredictable, go somewhere else!

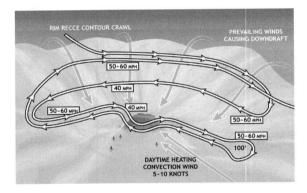

You can check out the strength of the downflowing air with the contour crawl. If there is any at the upper rim, stay in close to the headwall and keep the drop off. You will not have much visual reference to superimpose your normal horizon, so keep your scan moving.

If the downflowing air at the rim is not strong, do the next crawl about halfway down. make the altitude several hundred feet lower outbound and, when turning inbound, make any final adjustments to the halfway mark, which stops you descending too far outbound. If the wind is still moderate (meaning moderate turbulence), make the next pass an eye level pass by the intended landing zone.

Use the Figure 8, remaining close to the headwall inbound and outbound, remembering that the aircraft will try to climb inbound (with loss of airspeed), and vice versa. Always climb at least 100 feet when outbound so that you don't descend. Do an overshoot in the landing direction, no slower than about 20 kts. If you are happy, do the proper approach, always watching for wind shifts and keeping the drop off for as long as possible. Stay with the head- and sidewalls during departure for the maximum drop off and airflow conditions.

On takeoff, you may need to climb to avoid obstacles before heading for the exit.

Canyons

These are narrow steep-sided valleys that have had rock and surface material cut away from water erosion. They are the opposite of ridges. Creeks, rivers and streams may well be flowing out of them. If a river cuts down to some depth below the surrounding terrain, we are also dealing with a river bed.

Landing sites on the bottoms of valleys often have difficult access, and frequently leave no escape route once an approach has started. In this case, it's important to have safe power reserves before committing yourself. In any case, placing the aircraft downwind near to ground should be avoided, but if you have to, go low and slow when approaching downwind with a last minute turn into wind.

In snow, try landing with the sun behind you, as the aircraft shadow will give you a useful guide to the ground slope and surface and provide a focus for a sight picture approach. Some people use the landing light.

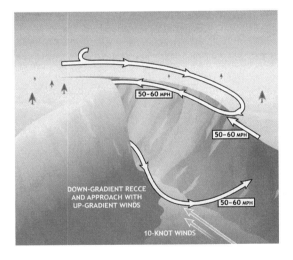

In all cases, the main types of wind will be valley winds, and those arising from convection and evening cooling (anabatic. katabatic, etc.) which can alter the direction of the valley winds. The winds will either be going up- or down-gradient. If you are forced to go up-gradient because the wind is down-gradient, be prepared for a drop off if things get marginal.

Start with a high recce, working down the gradient at the speed used for the contour crawl, checking for wires and other hazards. The taper of the rim can indicate the gradient at the bottom. If it is very gentle, you can use both directions. Keeping the same speed, fly along the rim up to the head, checking for hazards and markers, keeping the landing site in view if you have selected it.

© *Phil Croucher, 2016*

Find a wide spot to turn in, then descend from the headwaters down the gradient to the bottom of the canyon. When you see the marker for the landing site, slow down and get ready for the first overshoot.

Don't slow down too much! Use at least 40 knots in moderate winds and 20 knots in light winds. Having checked the landing site for suitability, climb up out of the canyon. You could have a go from the other direction if the bottom of the canyon has no gradient, otherwise reposition for the final approach.

At the landing site, place the tail rotor over the creek or stream (use a marker to help), with the nose pointed 45° into the gradient.

Taking off down-gradient is always safest, even if you have to go downwind (watch those performance charts!)

Narrow or Dead-End Valleys

A narrow valley is one where your radius of turn exceeds half the width of the canyon, so you are often better off flying on the downdraught side to keep you out of a worse situation if you have to turn round (you will actually be entering an updraught). However, you will also be subject to a tail wind that will increase the radius of turn.

There may be a very steep gradient at the bottom, possibly with a creek or stream flowing out. With narrow valleys, there is even less of a normal horizon, there is a funnel effect for the wind and fewer escape routes the more you go in. It is also easier to stretch cables across them!

For these reasons, it is best to go towards a valley exit and any low ground. If you have to fly up a valley, be aware of

the point of no return, and where the valley gradient ensures that a climb out will not be possible either.

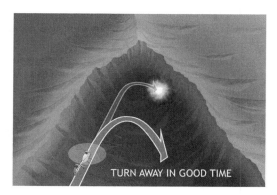

TURN AWAY IN GOOD TIME

Landing sites in the bottom of a valley often have difficult access and require you to be committed to landing from a longish way out because of the lack of escape routes. One good reason for having reserves of power, especially if the valley is at altitude!

Approaches to sites in valleys should therefore be treated with more caution. Think about the takeoff as well! That and the landing can be made into wind if it is along the axis of the valley, but if there is no wind, or it is across the valley, at least try to land and takeoff down the slope.

- Check the rim out at contour crawl speed, checking for up- or downflowing air.

- Check for wires and hazards such as trees or bottlenecks.

- Work your way across and down the lower mouth of the valley for landing sites (you won't be able to do this down-gradient).

- Work passes across the mouth of the valley, eyes level with the potential landing zone.

- Go up at least 100 feet on the outbound.

- Break away from the overshoot early enough to avoid getting trapped.

- Use a good rate of closure right to the landing site, keeping the disc loaded and avoiding any kind of flare, as you normal perceptions will be severely disrupted. Keep scanning!

- Land smoothly and carefully from a low hover to avoid running out of left pedal if the wind shifts.

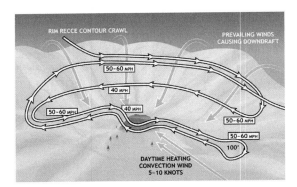

On the takeoff, do not pedal turn out of wind so the tail is stopped from hitting the sides (you will be going out the same way you came in). Instead, pick a marker and get into a high hover in a position that gives you some dropoff. As you get to the top of the tower, turn to the dropoff around the tail rotor. You will need to ease off the left pedal (in a Bell) in a right turn, then increase it again towards the dropoff, together with some cyclic movement.

Riverbeds

Start with a recce at a safe height, checking for the usual suspects, wires, trees, etc. Use the contour crawl along the upper banks, checking for up- and downdraughts, and potential landing sites. Use an eye level pass if possible to check the site out, try an overshoot, then an approach, using a distant marker to keep straight.

Typical airflow in a riverbed is shown below.

Cols (Saddles)

A col, or saddle, is a depression between two hills. A wide saddle has enough room for you to check it out on the upwind side, whereas you must pass through a narrow saddle to do an effective recce.

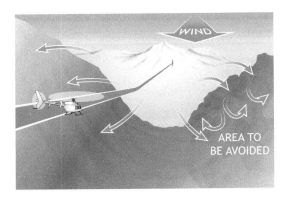

As saddles are below the peaks and ridges, the problems associated with them are mainly related to wind direction* and velocity, in terms of downdraughts and turbulence. The narrower the saddle, the more the wind will change.

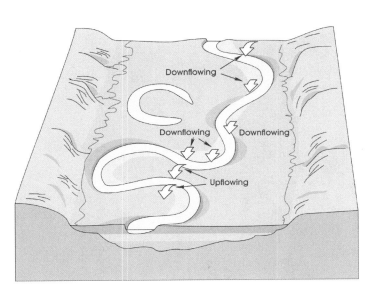

*In fact, you have a double venturi, so the effects of wind will be greater. In which case, beware of turbulence around any obstacles or rocky outcrops. You will need to combine the flying skills required for slope and valley flying, plus passage of a crest.

- You need a good height margin

- Fly along a slope and not along the axis

- Establish where you cannot make a a 180° turn

- Don't fly over sharp features

TAKEOFFS

Minimal time in the hover is also a good idea on takeoff. Not hovering at all is still not a good idea, because you want to be sure your skids are not caught in anything that might give you dynamic rollover. However, once about a foot clear of the ground, and you are sure you are not caught, pull in what power you have and rotate without delay towards downward sloping ground. Aim for speed and only reduce power when you can safely fly away.

To get off from high ground when the machine is heavy, and if you have a good enough surface down the hill, get light on the skids, hold the cyclic neutral and let the machine start to slide down.

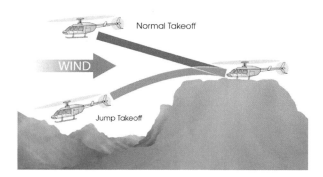

Then start a slight forward cyclic movement whilst waiting for the opportunity to pull enough collective to pull you off the ground.

Downwind
From an article by Matt Johnson.

Many instructors forbid downwind takeoffs, but they can be a useful weapon in your arsenal if they are handled properly with due regard for the aerodynamics involved and the performance aspects. For example, the bottoms of glaciers are usually troughs of downflowing air between two moraine walls, featuring descending air which you won't be able to climb into, because of the rising terrain ahead. Therefore you will have to depart downwind.

.In the picture below, the helicopters are using power according to the airspeed over their rotor blades. No 1 is using the most, with no airspeed, as might be expected. Nos 2 and 3 have 15 knots across the disc, and are using less power than No 1.

They are using the same power, as the rotor disc shouldn't care which way the wind is coming from (in practice, the downwind helicopter has dirty air coming the tail rotor, so a little more power is needed).

The problems arise when you try to take off in helicopter No 2. It loses the airspeed over the rotors almost immediately, and will not get it back again for a time, so you will need at least as much power as No 1 until you get past the zero airspeed zone.

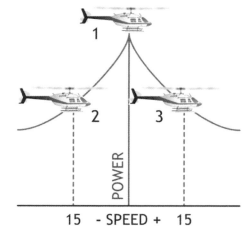

When taking off downwind, therefore, you must get used to having a high groundspeed with less airspeed.

You must also be lighter than usual to ensure you have enough power.

HELI SKIING

This is really a combination of mountain flying and playing around with snow, plus a couple of extras to watch out for, not least your own limitations and level of training, because you could be doing well over 40 precision landings a day, in a fully laden helicopter in tricky winds and weather, which is another thing to watch for, particularly the temperature and dew point, and how close they are together, which leads to fog and cloud, and poor visibility (check them in relation to the relative humidity, or how moist the air is). Another good one to check is icing, and where the freezing level is. So is the windspeed and direction. The usual practice is, if one pilot calls off for weather, everyone else does, too - there are many recorded services for winds and temperatures on the hills.

Do not:

- lose sight of your reference. Always turn towards it, and don't give one up until you see the next one. If you are not sure, do a vertical takeoff, then decide (this gets you out of the snow cloud, and keeps the tail rotor away from the clients).

- flare at touchdown - you will put the tail in the snow.

- commit to the spot before it is necessary, and you have done a power check.

- assume the wind, or the power. Check before committing yourself!

Do:

- Keep in touch with flight watch.

- Remember that the guide is not the Captain!

If you're flying more than one group, anticipate how long it will take to pull them all out with regard to how the weather is deteriorating. If need be, shuttle them to a place that can be used as a staging area as you pull the other groups out of the hills. Picking up at the base of a glacier with level terrain behind will take lots of power!

CONFINED AREAS

A confined area restricts your operations in some way. Most people think of them as clearings in forests, like this:

but could also include areas between buildings. Even a roof or an oil rig comes into the definition - they come in all shapes, sizes, and depths, in valley bottoms, on mountain tops, etc. In any case, you usually have to look down and in at a sharp angle to check out where you are going to land. You are likely to lose translational lift just as you begin your descent on a steep approach, or before descending on a vertical approach. You might have to descend into shadows down onto a landing site that is not visible underneath you. Descending below obstacle height, you could enter down flowing air coming over the trees. There are many reasons to make good judgments before committing yourself.

At higher elevations confined areas are usually easier to look at, get into, and out of, mainly because the trees are shorter, if there are any at all. It is therefore easier to do a close-in check at eye level. Approaches and departures can be more shallow, but the landing site might have just as many hazards. You could go to many different landing sites at a moments' notice, many of which may have been neglected for years, however well they might have been constructed in the first place.

Watch out for:

- recirculation

- windshear

- turbulence, which will be found downwind of most things

- tail rotor

Know the wind direction, but vary the flight path according to areas suitable for forced landings. A crosswind approach could be acceptable (if you're not too heavy) if you can approach over a clear area, but if it is from the left (in a Bell) be wary of LTE.

First of all, other things being equal (that is, no low cloud base, etc.), fly around the clearing at a suitably safe height (not below 500 feet, and especially not downwind) and try to hover OGE. If you can do that, you should be able to do so semi-IGE in the area. Most people starting out might take a look at 1000 feet, and look again much lower (say 300-500 feet so you don't lose sight of anything), but you can do both at 500 feet with experience.

For a more formal power check, at a safe height near the landing site, fly straight and level into wind at the best rate of climb speed. Note the power used. Then apply maximum power to see what is available. You are checking the distance between the bucket speed at the bottom of the power curve and the power available, as shown here:

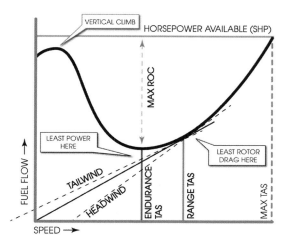

In a 300 CB, for example, a 6" spread in MAP could mean that a vertical descent is possible. Going the other way, if you have less than 1" (compare from a 2-foot hover), you may need a cushion takeoff, and plenty of room. If you have 1-2", you might get a normal takeoff. With 3" or

more, a towering takeoff, and you should be able to vertical out with more than 4".

Here are the figures for a Bell 206:

- **Departure**. Note the N_1 in a 2-foot hover and increase the power until a 5-minute takeoff power limit is reached (Torque, N_1 or TOT). The difference is your power margin.

N_1 Margin (%)	Takeoff
1	Cushion Creep
2	Shallow Climb
3	Steep Climb
4	Towering
5	Vertical

  N_1 is used because it provides a steadier reading.

- **Arrival**. From straight & level at 200 feet above the site at 40 kts, note the N_1 and increase the power until a 5-minute takeoff power limit is reached. The difference is the margin.

N_1 Margin (%)	Approach/Landing
4-5	Running Landing
6	Zero Speed
7	Normal to low hover
8	Normal to high hover
9	Steep to high hover
10	Vertical Approach

With Bell helicopters, 10% is more than enough to clear a confined area.

Sometimes, you may want to do an abbreviated procedure - customers very often don't want to know the gory details, so many pilots develop their own ways of estimating what power they might have at the landing site, while the final approach is being made.

For example, in a Bell 206, at a point where you can still turn away, flare slightly nose up (just above ETL), hold the flare until your speed falls through ETL and you will feel a slight shudder. At this point, level, apply the last of the power and note:

- Pedal position
- N_1
- TOT

To hover IGE it will take an extra:

- One inch of left pedal
- 12% N_1
- 35° TOT

To hover OGE will take an extra:

- 1.5 inches of left pedal
- 15% N_1
- 50° TOT

Another method mentioned previously is turn in at around 60 kts with the disc loaded, descending at around 250 fpm. The power used will give you a good idea of what you need in the hover, so you have an early chance to abort if you are using too much (you get to know with experience). This works, because 250 fpm reduces the thrust required to transition into the hover by about 15%, i.e. much the same as for ground effect and there should be minimal collective movement at the end.

If the air is moist, say after a shower, expect to increase Density Altitude by 1000 feet for every 10%. For a rough guide to your DA, set 1013.2 (29.92") on the altimeter, to get your Pressure Altitude. Then multiply the difference between ambient and ISA temperatures by 120 and add it to the PA.

On the way in, check for:

- **W**ind
- **O**bstacles
- **P**ath in
- **P**ath out
- **E**mergencies
- **R**econ (landing area):
    - Size*
    - Shape
    - Slope
    - Surround
    - Surface (snow covers stumps)
    - Sun (in your eyes)
    - Shadows (which could hide something nasty)

*The shadow of the helicopter will never be smaller than the machine itself, so if it fits in the clearing with a safety margin, so will you.

Always have your landing area in sight, or at least the boundary, and start the approach at around 60 kts (in a 206), with the disc loaded (early!), which gives you the best chance of reaching it if the engine fails, although risk management would suggest that you first worry about not hitting anything, or getting into vortex ring. Some people will approach their chosen spot directly from a fair height, but some fly level along the top of the trees, and go into an approach at a relatively late stage. However, the important thing is to *fly into the clearing*, and not to hover in, reducing speed as you get closer. For elongated holes in a crosswind, the solution is to face into wind then fly in sideways up the slot.

In a confined area, there will be a point beyond which you're committed, so don't go beyond it until you're sure. Pick a point to aim for where you know your tail and rotors will be clear, not too far towards the end, and fly in, as smoothly as possible, going over the lip to the clearing at around a fast walking pace, sort of horizontally (*flying, not hovering*). If you lose sight of the landing spot, go around. As for power checks, you will know very early on if you're running out. Keep an eye on the torque, and note any vibration in the pedals, which will give you a clue to the wind. As previously mentioned, descending at 250 fpm reduces the power required in a 206 by about 15%, so adding it to what you are currently using will give you an

If this page is a photocopy, it is not authorised!

idea of what you will get in the hover - just make sure it is nowhere near 100%!

The size of surrounding trees will give you a false illusion as to the size of the clearing, in that big trees will make it look smaller and *vice versa*. Watch their movement, too - your downwash could blow them away at the start, but they can bounce back into your rotors easily. Remember how you got in, and do the reverse on the way out.

A typical clearing will have stumps and slash all over the surface:

If you don't have logs to land on (and these produce their own problems when they are slippery), take off a cleanly as possible, to avoid your skids getting caught in something, being aware that tall trees will sway from your downwash. When landing, if there's room, try to move forward slightly as you touch the ground, as this will bring the tail up, away from the garbage. So as not to use pedals too much, you can use the cyclic to turn the machine if need be.

It is always a good idea to do a clearing turn before taking off, but often you cannot, so exercise extra caution if you think someone may be behind you. In a Bell, as you go out of a clearing, a little aft cyclic will produce a little extra lift, but don't expect the same from an AS 350, or you will clip the trees (in fact, with a Bell, you can work your way up a

little at a time by waiting for more wind, aft cyclic input, right pedal inputs and collective increases).

The key to getting out of a confined area is to get as much climb as possible going before you start to lose ground effect at approximately 1 rotor diameter. Hover low to start with, so you have more power available to initiate a good climb (use all the power you have). Once established in the (vertical) climb, the same power is used to maintain it, since it takes more to accelerate in the first place. You want a smooth collective pull to 91-95% torque, so you have some room for torque spikes.

Tip: If you are taking off over grass, notice where it stops being blown by the downwash. Assuming you set the correct pitch for departure, that will be the point at which you reach ETL. This is very useful when departing out of confined areas.

The speed of the wind is not so much of a factor as how turbulent it is. Turbulence can be created by the shape of the mountain or the nature of the weather system. Always consider your altitude. The aircraft will not be as responsive in thin air, so it may not be as safe to land at a higher peak on the same day.

Because the weather changes so much, there are no hard and fast rules - very often, you have to make it up as you go along. Other than that, the only fixed item on the agenda is the shape of the ground you are working over. Round, smooth ground is much easier to work with. If there is a cliff is on the downwind side of a peak, be careful of the diving curl behind it. Landing on a shoulder behind a peak will also be difficult, and made worse by a windy day, as even a smooth wind will now be disturbed by the peak.

Before landing somewhere, consider the takeoff - try to remember markers on the way in so you can pick them up again on the way out.

SEARCH & RESCUE/HOISTING

6

oisting may be used for rescuing people in distress from land or water, or for the routine transfer of people and/or cargo to or from rigs or vessels, or even dropping off firefighters (you won't normally be required to pick them up again). As a method of rescue it is a procedure that matured during the Vietnam war. It calls for a high degree of crew co-operation and trust, since the pilot relies on the Hoist or Winch Operator for safe positioning, as he cannot see the target from the overhead. This, in turn, requires a concise and accurate *con* (continuous instructions in the form of a commentary) and a thorough knowledge of standard procedures and phraseology, plus the willingness to abandon a job if necessary, to keep within acceptable risk (although standardisation is officially a Good Thing, the nature of rescue work requires you to think on your feet, so some latitude is permitted to get the job done).

In view of the above, the following general points apply:

- As humans are lifted, there are special considerations - Hoisting is a Class D External Load. In an emergency, this is the priority for crew actions:

  - The safety of the helicopter and crew

  - The safety of the people being hoisted

- Always ensuring that you (or the watching public) don't become victims as well, *use the lowest risk method first*, escalating it only if it doesn't work. Crews should be able to turn down missions if they cannot be performed safely

- Response time should be as short as possible, consistent with proper planning and flight safety

- The helicopter must be capable of sustaining a critical power unit failure with the remaining engine(s) at the appropriate power setting, without hazard to the suspended person(s)/cargo, third parties, or property. Minimal time should be spent in the hover, because of the possible inability to maintain it if an engine fails. Short sorties are preferred, but this depends on the nature of the emergency. The time spent on site will reduce with your hover height, but the effects of downwash will be greater. On the other hand, the higher the height selected for the hover, the greater the margin of safety there is if an engine fails

- Since every case is different, a degree of flexibility is allowed. Although the decision to start a flight is the PIC's, once on site, the PIC and HO (Hoist Operator) consult between themselves, and either may terminate the flight if the risk becomes unacceptable. The chain of command is PIC - HO - Rescueman (if present)

- There should be enough seating so your passengers don't have to use the floor

- Pilot permission must be gained before opening the cargo door in flight

- The rear seat crew should have clear visors down, unless it compromises their safety, such as during night operations and in heavy sea spray

- HISLs (*High Intensity Strobe Lights*) should be switched off before starting

- *At no time* should the hoist cable be tied to the vessel or platform

- As with any slung load, watch out for static electricity. The responsibility for earthing the winch cable normally rests with the vessel

- *Do not lose the casualty!* There may be an element of liability on your part if the casualty is dropped and dies as a result. However, to repeat: In an emergency, this is the priority for crew actions:

  - The safety of the helicopter and crew

  - The safety of the people being hoisted

DEFINITIONS

These may vary between companies:

- *WO/HO* - Winch Operator/Hoist Operator. The senior guy in back, normally a Para-jumper or Paramedic, in charge of the cabin and RC. Pretty much the man in charge once on task

- *RC* - non-specific rear crewmember, referred to as *One* or *Two* (there should be two, but you can use one with a suitably qualified person on site). The term includes the Winchman or Rescueman (the one that gets wet at the end of a wire)

- *FC* - the pilots, normally used in derogatory fashion

- *Standby position* - sometimes called the *Rest Position*, this is a hover alongside the vessel or site at a comfortable height (50-100 ft), on a line roughly 45° from it, so that, in a forced landing, the helicopter would clear the area. It is also from where the helicopter will move over the site and to which it will return as each hoist cycle is completed, thus removing downwash and noise from the area (around two rotor widths keeps rotor wash away from the site). While in the Rest Position, the pilot can look down and to the right and use the scene as a hover reference point, and a short briefing concerning power requirements can be given. Factors to be considered include:

    - *Length of Cable*

    - *Wind*. Its effect on the helicopter, cable position, and turbulence

    - *Visual Reference*. Discuss any action in case it is lost (i.e. in blowing snow). The slow roll or pitch of a vessel may make it a difficult reference, but the waterline is an excellent one

    - *Power Requirements*. Single-engine capability - engine power available - check performance charts to determine HOGE/HIGE. For training, a 15% or greater power reserve is highly desirable

    - *Ditching / Forced Landing*. When either is likely after an engine malfunction, discuss how to move clear of the ground party/deck

    - *Hover Height*. 10-20 feet is suitable for an S76, 20-25 feet for a 212, and 30-40 feet for the S61 and AS332

- *Hoist Cycle*. One down-and-up cycle of the hoist hook. The number used should be recorded in the Tech Log

- *HHO Site*. An area in which a helicopter performs a hoist transfer. Ideally, it should be at least 10 feet square, provide a level, non-slip surface free from spray and at least a 20-foot horizontal rotor clearance at all hover heights

- *HHO Passenger*. A person to be transferred with a helicopter hoist

- *LKP*. Last Known Position

Standard Terminology is covered separately, below.

LEVELS OF SAR SERVICE

Level 1

Using a dedicated helicopter, equipped for day, night, land, boat and open water hoisting, with a crew of two pilots, hoist operator and rescue people.

Level 2

Using a helicopter equipped and crewed for various levels of SAR, as determined by contract, with a crew of two pilots, hoist operator and rescue people.

Level 3

Using a helicopter with other primary duties, so the hoist is not normally fitted, which means that the response time may be over one hour, and the operating envelope will be significantly reduced:

- Over land, the suggested hoist zone will be restricted, say:

    - Hoisting area 4.5 m in diameter, with no obstacles

    - A secondary area 15 m in diameter with no obstacles over 3 m

    - A tertiary area 30 m in diameter with no obstacle higher than 15 m

- When offshore, the minimum vessel size should be 150 feet

- Open water or night hoisting should not be permitted

- Weather minima 300' ceiling and 1 mile visibility

- Max wind 50 Kts

- Fuel reserves should allow for 30 minutes over the site in a hover

- Active hoisting (i.e. lowering a qualified hoistman to prepare the hoist scene and/or the people being rescued) should not happen

- The hoist operator's duties should be limited to lowering the hoist to a prepared site, and recovering the hoist on board

PLANNING

You will mostly not be able to do much of this until you have seen the location, so, to minimize the time in the hover, the Pre-Hoist Check, site inspection, last minute planning briefing and discussion of emergency procedures could be done while orbiting it.

WEATHER & OTHER CONDITIONS

Normal weather limits should apply, but common sense prevails if you are picking someone out of the soup. The suggested maximum vessel or platform motion should be:

- Pitch - 8°

- Roll - 15°

- Heave - 10°

EQUIPMENT

Winch Cable

As with slinging, these items will render a cable unserviceable:

- visible bends

- crushing

- broken strands

- corrosion

Minor *birdcaging* (strand separations) can be caused by repetitive use of the first few metres, which can usually be eliminated by full extension with the maximum load on. If birdcaging cannot be smoothed out, the cable should be rejected.

Utility Hoist

The primary tool for rescues, which can be electrically or hydraulically operated. A hoist can generally carry up to 600 lbs (two people).

Rescue Sling (Horse Collar)

This is the basic hoisting tool for an uninjured survivor capable of self-help, assuming it is put on properly. It is designed for when pressures on the upper body will not create further difficulties, since it butts up under the arms. There is a webbing handle at the back, which gives the hoist operator a handhold for assisting the person into the helicopter. It can also be used to lower people to a site.

Hoist Harness (Double Lift Harness)

A body harness with a specially designed fitting for use with the hoist hook. It allows whoever is being hoisted to have free use of their hands. With a suitably qualified HO, it can be used to lower and recover trained personnel.

DOUBLE LIFT HARNESS/RESCUE STROP

This combination is used for patients who have difficulty getting into a Rescue Strop. The Rescueman is lowered in the Double Lift Harness with the Rescue Strop attached to the hook, where he places the strop around the evacuee. When secured, they are hoisted together.

Rescue Net (Billy Pugh)

The "Billy Pugh" net is used to hoist partially incapacitated survivors from water. It is not as effective as the Double Lift, but is useful with only one HO on board.

Stokes Litter

This is a rigid, protective wire mesh stretcher with a hoist bridle which allows it to be delivered and recovered with the hoist. It provides support and protection for casualties with torso injuries. It is always used with a "Y" strap, breaklink and guideline attached.

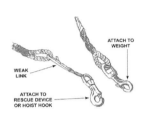

- The *Y Strap* holds the stretcher parallel to the body of the helicopter and keeps it from spinning.

- The *Breaklink* (left) is a loop of 80 lb test cord, which connects the "Y" strap to the guideline. Its should break if the guideline gets snagged.

- The *Guideline* (attached to the "Y" strap) stabilizes the stretcher. It is a separate line used by the Rescueman to stop it spinning.

Additional Miscellaneous Gear

The checklist should include:

- Hoist control (pistol grip)
- HO intercom with hot mike (a No-Go item)
- Cable cutters with stowage and a lanyard
- Cutting tool (knife) with stowage and a lanyard
- Rescue Strop (if required)
- Hoist-capable stretcher with restraint straps, lifting bridle, "Y" strap (if required)
- Guideline (if required)
- Restraint harness with safety lanyard
- Safety lanyard attachment point
- HO helmet & gloves (durable, well fitting with good dexterity)
- Hypothermia suits & jackets

The HO must decide whether or not to wear flotation gear over or under the restraint harness. If an emergency occurs and he has to go out with the harness on, it should be over the harness. Otherwise, it may go underneath.

SEARCH PATTERNS

These are orderly methods of searching an area, with different procedures for various circumstances. Search intensity is a function of altitude, airspeed and track spacing. Altitudes below 800 feet are actually only effective below 90 kts with a very small target and area to search. Generally speaking, 800 feet with a track spacing of 3 miles is good enough.

ELT Aural Homing

In straight and level flight, record the position the instant a signal is received. Then turn onto a cardinal heading (True) and continue until the signal fades out. Record that position. Reverse course to a point halfway between signal reception and fade-out (between the two positions) and turn 90° left or right. Continue until the signal fades out. Record the position and reverse course again. Continue until signal fade out and record the position.

Reverse course again to a point halfway between the two positions and conduct a visual search (using the Expanding Square or Sector Search).

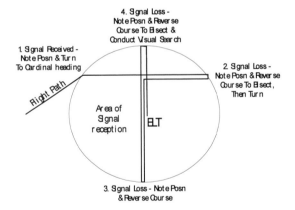

Expanding Square

As the name implies, this comprises a square in which the individual legs are flown further and further (expanding) from the LKP (*Last Known Position*). The initial leg would be flown on a cardinal track (true) and be 1 nm long.

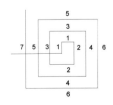

Track Crawl

This is based on the track of the missing aircraft, and used when the time it went missing is not known for certain. The search area is along the flight-planned route and out to 15 miles either side, starting 15 miles beyond the LKP and ending 15 miles beyond the destination, unless it is obvious that the aircraft would not have proceeded beyond this point. Just fly parallel tracks within that area (see below).

Parallel Search

This can be utilized in a track crawl (above) where the axis of the search is along the vessel's original track, with successive legs offset alternately to the left and right. Search intensity is a function of altitude, speed and track spacing. Subsequent searches of the same area should be flown at 90° to the original.

Sector Search

This pattern is used when the position of distress is known within close limits and the area to be searched is not extensive. It is simpler to execute, provides greater navigational accuracy, and is more flexible than the expanding square but, more importantly, the track spacing is small near the centre point and larger at the extremities, resulting in an increased probability of detection near the centre, which is the most likely position. If a drifting datum marker has been deployed, as the aircraft passes over it, the datum point may be re-oriented. This adjusts the search area for the drift of the target.

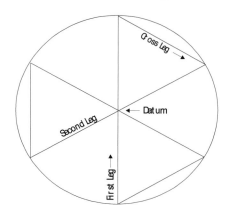

Coast Crawl

The coast crawl is fairly straightforward. It starts along a shoreline with successive lines parallel to the coast at an increasing track spacing (track spacing must be estimated by eye). Tasking for a coast crawl will normally include a Start and a Stop point and a maximum distance out from the shore.

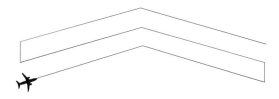

Contour Search

In mountainous areas, this is the only suitable search pattern. It is also hazardous, and should only be assigned under the following conditions:

- the aircraft must be highly manoeuvrable, of low speed with small turning radius and adequate power reserve

- the crew must be experienced in mountains, well briefed with suitable maps

- only one aircraft should be assigned to an area. Multi-unit contour searches should be done only by ground search teams

COMMUNICATION & PROTOCOLS

Hoisting requires a high degree of cooperation and mutual trust between crew members. The rescueman relies upon the HO and PIC for his safe recovery, while the HO relies upon the PIC to maintain position above the target. The PIC, in turn, relies upon the HO and rescueman for directions and operational advice (as he cannot see the target), and correct management of the situation at the end of the hoist cable. Due to the high noise environment and the need for the HO to have a live mike during the operation, communications must be clear, concise and unambiguous.

Flight Following

It may not always be possible to keep flight following updated, especially in the hover. Although you are not expected that you will stop halfway through a rescue to climb up and make a report, don't forget that overdue action will be taken if you are not heard from. Also, it is good airmanship to remain in contact with *somebody*, whenever possible, even lowly fishing vessels, as they at least have common marine channels (Channel 16, for example).

The Commentary

The commentary, or *con*, given by the HO (*Hoist Operator*) is highly important, and should use standard terminology, because in the final stages you lose sight of the survivor, and need the HO to guide you to the overhead, keep you there, tell you what's going on and keep you clear of obstructions while you reduce speed to the hover and descend to around 30 feet. So important is the con, that radios should be turned down or switched off for the lift.

Line And Hover Corrections

These are passed in units, which are not specific measurements, but *gauges of distance* that allow the pilot to accurately judge the relative rate of closure - that is, they are not specific measurements, but around 6 feet long - it's the *rate* of countdown that counts. Alterations to the approach direction are *line corrections*. Horizontal distance is reported in *units*

Order	Meaning
FORWARD x UNITS	Move in the direction stated maintaining into-wind heading. Distance should be reported continually so the pilot can select appropriate speed. Use clock code to refine direction
FORWARD AND LEFT x UNITS	
FORWARD AND RIGHT x UNITS	
LINE GOOD	The direction in which the aircraft is moving will intercept the target and is to be maintained

Speed Corrections

These are not in increments, as ground- and airspeed bear little relation to track.

Order	Meaning
REDUCE SPEED	Reduce speed
INCREASE SPEED	Increase speed
SPEED GOOD	Speed is appropriate for that stage
5 4 3 2 1	Prepare to adopt the hover (used in the final 5 units)
STEADY	Take up a stationary hover

Height Corrections

Vertical distance is reported in increments of feet, because you may need to protect the winchman.

Order	Meaning
UP X FEET	Climb the number of feet stated
DOWN X FEET	Descend the number of feet stated
HEIGHT GOOD	When used during the approach, the aircraft is at an appropriate height for that stage of the approach
	In the hover, the report indicates that the aircraft's height is correct and should be maintained
UP, UP, UP	Climb Immediately

Standard Terminology

Directions are given relative to the fore & axis of the helicopter (not the vessel)

Command	Meaning
UP	raise the hover height
DOWN	lower the hover height
LEFT	move to the left
RIGHT	move to the right
CLEAR	no obstruction within 25 ft of the rotor disc or 10 ft vertically of the aircraft belly or rotor blades
STEADY	maintain relative position
DOOR OPEN	door is fully open and hoist actuation is about to begin
BOOM EXTENDING	boom is being extended laterally from its stowed position
CABLE GOING OUT	HO is extending cable from the hoist cable drum
CLEAR THE DECK	the person is off of the deck and on the way up
CABLE COMING IN	HO is retracting cable onto the hoist cable drum
MAN IS ABOARD	the person is off the hoist hook and secure in the aircraft
BOOM RETRACTING	the boom is being retracted into its stowed position
DOOR CLOSED	door is fully closed, latched and secured for forward flight

Hold is an ADVISORY call given by the HO to indicate that time is required to retrieve the hoist/casualty into the helicopter or to get the casualty safely on the deck.

Emergency Terminology

Repeating a command 3 times indicates imminent danger, requiring *immediate* action:

	Phrase	Situation	Action
1	UP UP UP	Imminent danger below	Climb Immediately
2	RIGHT RIGHT RIGHT	Imminent danger left	Move rapidly right
3	LEFT LEFT LEFT	Imminent danger right	Move rapidly left
4	CUT CUT CUT*	Given by PIC to advise PM/HO that aircraft is in danger (power loss, etc.)	PM and HO will cut cable electrically or manually
5	ABORT ABORT ABORT	Given by PF to indicate a serious problem that requires immediate ceasing of operations with minimal risk to the casualty/survivor but possibly at the expense of the cable	HO will retrieve the cable/casualty into the cabin or return the casualty/survivor to the ground/deck/water and call "CUT CABLE" when he is "safe"... then "Clear to fly" once the cable is clear or inboard
6	EMERGENCY BREAKAWAY EMERGENCY BREAKAWAY EMERGENCY BREAKAWAY	Given by PF to indicate an emergency which requires that he depart the hover immediately regardless of the risk to the casualty/survivor	HO will retrieve the cable or call "CUT CABLE" if retrieving the cable is not practical. He may be able to reduce the likelihood of injury to the casualty by delaying the "CUT CABLE" call for a second or two until the casualty can drop from a lower height or into a friendlier environment
7	DITCH DITCH DITCH	From the PF, indicates that an emergency requires immediate ditching regardless of the risk to the casualty/survivor	HO will call "CUT CUT CUT" immediately

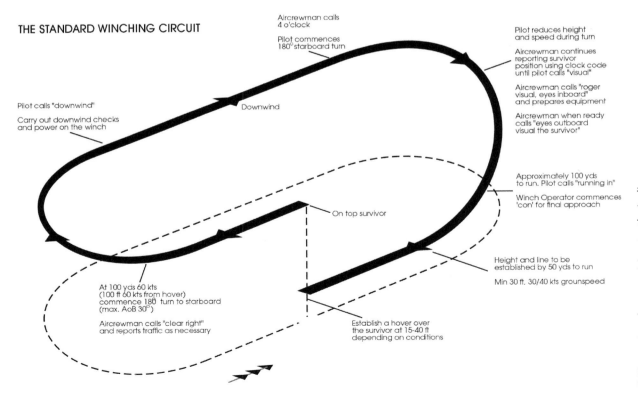

THE STANDARD WINCHING CIRCUIT

Aircrewman calls 4 o'clock

Pilot commences 180° starboard turn

Pilot reduces height and speed during turn

Aircrewman continues reporting survivor position using clock code until pilot calls "visual"

Aircrewman calls "roger visual, eyes inboard" and prepares equipment

Aircrewman when ready calls "eyes outboard visual the survivor"

Approximately 100 yds to run. Pilot calls "running in"

Winch Operator commences 'con' for final approach

Pilot calls "downwind"

Carry out downwind checks and power on the winch

Downwind

On top survivor

Height and line to be established by 50 yds to run

Min 30 ft, 30/40 kts groundspeed

At 100 yds 60 kts (100 ft 60 kts from hover) commence 180° turn to starboard (max. AoB 30°)

Aircrewman calls "clear right" and reports traffic as necessary

Establish a hover over the survivor at 15-40 ft depending on conditions

STANDARD CIRCUIT

The aim is to ensure that the survivor can be seen from the helicopter at all times, as a body in the water is a very small target. An accurate con is most important, as is smooth control of the aircraft. The complete circuit is flown at up to 200 feet, at 50-70 knots, and is used for basic training, but may be adapted as necessary. The preparations for the rescue are completed by the crewman during the circuit.

Stage 1

The aircraft overflies the survivor on the into-wind heading, maybe marking the position with smoke, as long as there is no hazard (i.e. fuel on the surface).

Stage 2

At around 100 yds upwind of the survivor (or at 100 ft, 60 kts if transitioning from a hover), start a 180° turn to starboard (20° maximum bank), steadying on the downwind heading.

Stage 3

Pilot calls "DOWNWIND" and carries out the downwind checks which are the Abbreviated Landing Checks:

"LANDING GEAR - WHEELBRAKES - TAILWHEEL - HARNESSES"

The aircrewman calls "WINCH CONTROL TO CREW". The aircraft continues on the downwind heading until the point where the survivor is in the 4 o'clock position, when a second starboard turn of 180° is initiated.

Stage 4

During this second turn height and speed are reduced. The position of the survivor is reported by the aircrewman using the clock code until the pilot calls visual with the survivor. The aircraft steadies on the into wind heading downwind of the survivor. Once the pilot is 'visual' the aircrewman calls:

"ROGER VISUAL EYES INBOARD"

in order to prepare equipment and check the winchman (if using the double lift method. Once complete the aircrewman calls:

"EYES OUTBOARD AND VISUAL".

The pilot then calls:

"RUNNING IN"

at around 100 yards from the survivor a final approach to the survivor under guidance from the winch operator.

Stage 5

With around 50 yards to run, the aircraft should be at least 30 feet agl/amsl and 30-40 knots. A steady hover should be established directly over the survivor at 15-40 feet, depending on the wind conditions and the state of the surface from which the survivor is to be lifted (rough sea, ground, cliff or vessel at sea).

Actions & Commentary

Crew Member	Action	Commentary
Pilot	On top of survivor into wind	On top survivor NOW NOW NOW
Winch Operator	Visually sights survivor	"Roger"
Winch Operator	Reports survivor's position	"Survivor 6 o'clock 100 yds
	100 yds astern a/c checks clear right	Clear right or report conflicting traffic as appropriate.
Pilot	Commences 30 Deg turn right onto downwind hdg	"Roger turning"
Pilot	Steadies on d/wind hdg	"Downwind"
Winch Operator		"Roger downwind. Winch Control to crew"
Pilot	Selects rescue hoist to crew.	"You have winch control"
Winch Operator	Lowers winch to attach equipment	"Roger I have winch control"

Crew Member	Action	Commentary
Winch Operator	Maintains visual contact with the survivor reporting when the survivor is in the right 4 o'clock position	"Survivor 4 o'clock"
Pilot	Checks clear right Commences 30 Deg right turn onto into wind heading.	"Roger turning".
Winch Operator	Continues to report the position of the survivor using the clockcode until the pilot regains visual contact.	"3 o'clock 2 o'clock etc."
Pilot	On regaining visual contact.	"Visual"
Winch Operator	Lowers the rescue hoist and attaches the appropriate rescue equipment.	"Roger visual eyes inboard preparing gear."
Pilot	On receipt of the Winch Operator's Report	"Roger equipment checked"
Winch Operator	If employing the Double Lift moves the winchman to the doorway in readiness. Regains visual contact with the survivor.	"Eyes outboard Visual"
Pilot	Continues descent requests commentary when required (normally at 100 yds).	"Running in"
Winch Operator	Guides the a/c using the standard commentary to the on top position	"Roger Running in"
Pilot	Gradually brings a/c to winching height and establishes line for final run in. Height and line to be established by 50 yds to run.	
Winch Operator	Continues to guide a/c using standard con whilst keeping the pilot informed about the progress of the operation and any potential hazards or obstructions	i.e: 50 units to run height good reduce speed 40 units etc
Pilot	Continues final approach as instructed by winch operator.	
Pilot	Maintains hover relative to ground water or ship as instructed by the Winch Operator.	
Winch Operator	Provides a continuous commentary until the operation is complete.	
Winch Operator	Ensure survivors/passengers are secured. Secures the rescue hoist Checks that the aircraft is clear.	Winch housed power off the winch clear right above and behind. Check clear left- to transition Clear to transition.
Pilot	Selects rescue hoist to off	
Winch operator	Secures the cargo door	

THE HOIST

Single Lift

This is the normal way of transferring people, but it should only be used as a last resort for rescue purposes, which would normally require the double lift, below. The strop must touch the water before it is touched by the survivor in order to discharge static electricity.

Enough winch wire should be paid out so that the survivor can compensate for the wave action whilst putting it on, but too much can be a snagging hazard. That is, there should be as little slack as possible.

Double Lift

The double lift is the standard rescue method if you have the equipment and crew, because the winchman and survivor can be lifted together. Static is discharged with a zap lead. Duties of the winchman and winch operator are allocated by the aircraft captain.

On the final run-in the winchman should be lowered so that, below approximately sea state 3, his knees are in the water with 15 yards to run, helping to stabilise him in the direction that he is travelling. However, in rough sea states, the winchman should be kept clear of the water until reaching the survivor. Because the winchman is connected directly to the winch wire, a fair amount will need to be paid out for him to work freely and to stop being pulled out of the sea in the wave troughs, but this must be carefully regulated to prevent a snagging hazard.

Once clear of the water, the survivor and winchman can be recovered as normal.

Double Strop Hypothermic Lift

To reduce the risk of cardiac complications after prolonged immersion in cold water, you can recover the survivor in the seated position using two strops. A static discharge lead should be attached to the winch hook followed by the two 8- foot rescue strops in the closed loop configuration and the double lift harness.

The winchman should be positioned near the survivor in the water. The first strop should be attached around the torso beneath the arms, with the second strop under the knees. Both should be tightened before lifting the survivor clear of the water. On arrival at the cargo doorway the winchman should remain outboard of the survivor so the WO can assist in the recovery.

Stretcher Transfers (SAR)

Stretchers can be used for injured or unconscious people. The winchman is lowered to the deck or ground before the operation begins and is then responsible for the general conduct of the stretcher transfer.

If a Doctor is available and weight limitations permit, he should be carried as well as the winchman and be lowered to the transferring ship to ensure the correct stowage of the patient.

If the stretcher cannot be brought onboard, it must be handheld and the airspeed kept below 30 knots

Practice runs should be carried out with a dummy!

FROM A SHIP

The winchman is transferred to the deck and the stretcher is lowered to him, where he guides it clear of obstructions, unhooking it when it is on deck. Alternatively, the winchman, in a Double Lift harness, may be lowered with the stretcher. The aircraft then moves clear to Port and remains in a hover relative to the ship.

The winchman briefs the ship's personnel, and the patient is transferred to the stretcher under the winchman's supervision. The patient's arms must be left free outside the stretcher harness if possible, to free himself in emergency. When the winchman is satisfied that the patient is correctly secured in the stretcher, the aircraft is called in from the waiting position.

The aircraft makes its approach and the hook is lowered to the deck to discharge static electricity. The hook must NOT be touched until this has been done and great care must be taken to keep the hook clear of guard rails and other obstructions.

The winchman attaches the hook to the stretcher and his own harness. He gives the affirmative signal to the Winch Operator when ready for the lift. The winch wire is plumbed and the stretcher is raised, with the winchman guiding the stretcher clear of obstructions. When the stretcher is clear, the aircraft moves to one side for the remainder of the lift. Ideally the patient should arrive at the cargo door with his head in the forward position.

TO A SHIP

Stretcher transfers are normally made to hospitals or sick bays ashore, or to ships with good medical facilities where a landing should be possible. The winchman is transferred to the receiving ship to supervise reception of the patient.

When the aircraft is called in for a transfer to ship, or there is no place to land on shore, the pilot may operate the winch as required by the WO whilst the stretcher is manhandled outboard. The winch operator then resumes control of the winch to complete the transfer.

DOUBLE LIFT

The winchman and stretcher will be raised to door level, with the stretcher closest to the door and the head held next to the winchman's right leg.

When the winchman can grasp part of the structure of the aircraft, the winch operator assists him to rotate if necessary. The lift is then completed to allow the winchman to be brought inboard. Winch operator and winchman then manhandle the stretcher inboard, the pilot having winch control.

Downwind Checks

Once established downwind, the PF calls for any checks remaining (i.e. 'Floats armed at 55 knots') and asks the PM to coordinate with the HO and call out and action the Pre-Hoist items on the Hoist Checklist.

Final Approach

On finals, the pilot should call for final checklist items to be actioned, transition to a hover in the rest position, at which point it is decided if the hoist mission is a go.

Who Has The Con?

Here, the HO has the con and directs the PF to the hoist site. The PF, PM and HO maintain an obstruction watch:

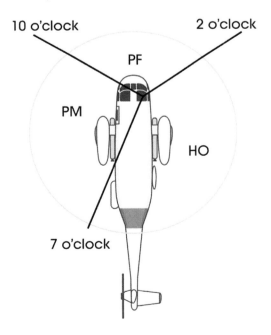

Return To The Rest Position

Once the mission is complete, the HO will return the PF to the rest position, with the PF reassuming the con. During this, the HO secures the person being hoisted in an appropriate place in the helicopter and prepare for the After Hoist Checks.

After Hoist Checks

After whoever being hoisted is secured onboard, the After Hoist Checks are read off by the PM and actioned by the PM and HO. On their completion (doors closed etc.) the PF may rotate and transition to forward flight.

EMERGENCIES

Hoisting is by nature hazardous - in the hover, the helicopter may be near maximum power, and the hoist wire may become snagged. In all cases, in an emergency, any load below the helicopter should be shed, and the dangling hoist wire removed (e.g. cut the cable).

Just to remind you, this is the priority for crew actions:

- The safety of the helicopter and crew
- The safety of the people being hoisted

Engine Failure

For practical purposes, after an engine failure, the PF must call CUT CUT CUT, then ROTATE. Due to the heights at which hoists are conducted, it should be possible to cutaway the person and allow him to fall back to the deck without too much further injury.

If hoisting from a stable vessel, and safe rotation over the water is possible (due to enough deck height and good obstacle clearance), the PF may instead call out ABORT ABORT ABORT and elect to continue with the hoist during the Flyaway. However, if the person on the hoist comes in contact with the water, the HO will have a discretionary call to make as to whether to cutaway or continue hoisting. If flight cannot be maintained, upon hearing DITCH DITCH DITCH, the HO and PM must cutaway the person on the hoist IMMEDIATELY.

Loss Of Reference

If the PF experiences loss of reference, his ability to maintain a steady hover or provide his own obstacle clearance is jeopardized.

The PF must immediately call "Lost Reference" and the HO must give directions at once to stabilize the aircraft. If this is not possible, the PF must rotate immediately to regain control.

If this occurs part way through a hoist, this will complicate matters, but remember the priorities above. If control cannot be established immediately through the HO's directions, the flyaway must follow without delay and the PF may opt to call for the person on the hoist to be cutaway if there is no danger to the aircraft.

Hoist Freeze

This is defined as any stoppage of the up or down action of the hoist. Possible causes include:

- Power Failure
- Mechanical Failure
- Pendant Failure
- Cable Drum Malfunction

Should the normal electrical hoist operating switch malfunction then down or up selection can still be made with the manual hydraulic buttons on the outside of the winch motor.

If the hoist fails during a transfer, the following should be considered:

- If the amount of cable already out permits, it may be possible to bring the aircraft down to a lower hover height and deposit the person on the deck
- The vessel could launch its rescue craft, into which you could lower the passenger

If neither are feasible, make a low and slow hover to land the person on the nearest Helideck.

Cable Runaway In/Out

This refers to an uncommanded actuation of the cable extend/retract mechanism. The following should occur:

- HO calls "RUNAWAY HOIST"
- PM switches HOIST PWR to OFF
- HO selects reverse control

Once HOIST PWR is OFF, the PF may opt to attempt a hoist with his own collective controls. If so (with agreement from the PIC, the PF may command the HO to disconnect Pendant Control Assembly and ask the PM to momentarily turn HOIST PWR ON. If there is no cable runaway, continue hoisting with HO advising the pilot when to raise or lower the cable with collective controls. If runaway continues, turn HOIST PWR to OFF and abort mission. If runaway is not controlled by the

above, the PF must lower the person being hoisted back to the deck after instructions from the HO.

Once the person is off of the hook, the HO should recover the cable by hand, return to the Rest position and carry on with After Hoist Checklist.

Ditching

The PF must announce his intentions to the PM and HO by saying DITCH DITCH DITCH. Both HO and PM must actuate their respective CABLE CUT switches, so the person being lifted becomes the person being dropped. When the CUTAWAY is completed, the HO must attempt to brace for landing by sitting straight and upright (if in an aft-facing seat) or sitting doubled over in a forward-facing seat. Before touchdown, the PF should announce BRACE BRACE BRACE to warn all on board to brace for contact with the water.

Once the engine and rotors have stopped, continue with the post-ditching procedures.

Load Swing/Spin

If the hoist begins to swing, the HO can dampen the effect by:

- stopping hoist movement
- pushing or pulling against the swing, keeping the pilot advised
- continuing with caution if the swing action is small

Caution: An unchecked swing may result in serious injury to the person on the hoist or impact with the aircraft.

Intercom Failure

This is usually the result of an unserviceable MIC switch and can generally be overcome by switching to another intercom cord or changing headsets. Hand signals should be used if the intercom cannot be restored. If additional crewmembers are on board, they should help keep the flow of information going.

Note: Hoisting without proper communications between the pilot and HO incurs significant additional risk. This should only be done to complete the immediate evolution, then the operation should be suspended until proper communication can be restored.

Hand signals for emergency use are accepted as follows:

Signal	Meaning
Arm extended, movement up & down with palm down	move aircraft down
Arm extended, movement up & down with palm up	move aircraft up
Arm extended with hand closed in fist	stop & steady aircraft
Arm extended with hand closed & thumb up	affirmative (Yes)
Arm extended with hand closed & thumb down	negative (No)
Hoist hit repeatedly with one hand & thumb down on other	hoist U/S
Arm horizontal with index finger extended and pointing left	move left
Arm horizontal with index finger extended and pointing right	move right
Arm horizontal with index finger extended, pointing forward	move forward
Arm extended by side with palm open, finger pointing down	hoist moving out from stowed position
Arm crossed over chest with palm open and fingers pointing up	hoist moving in from extended position
Both arms crossed over chest with hands closed	hoist stowed and hoist operation completed
Slashing movement with hand across throat	CUT CABLE or cable has been cut
Hand with index finger pointing down with spiral motion	cable going down
Hand with index finger pointing up with spiral motion	cable coming up

Fouled Hook

If the hook gets entangled or snagged, it must be cleared immediately, either by running out further cable or through hand signals from the HO to crewmen on the deck.

Warning! This is an extremely serious situation and could cause the aircraft to crash!

Note: Before the hoist, the briefing to the person being lifted should include a warning for him to guard the hook and rescue strop and not to let it touch the deck or let it out of his sight. Should it become snagged, he must be briefed that he must make every effort to free the cable or the HO may be forced to cut it.

The HO must inform the pilot of the situation as soon as possible. If it is not possible to clear the cable or hook immediately, and the aircraft is in danger, the CABLE CUT is to be activated on command of the pilot, especially from a heaving vessel.

Warning! The HO should not place any tension on the cable by manoeuvring the aircraft in an attempt to free the fouled hook!

SAFETY MATTERS

Restraint Harness

Restraint harnesses and safety lanyards should always be worn near open doors.

Safety Lanyard

This should be adjusted so that no more than one third of the body projects beyond the door opening. The HO must be able to reach essential objects or tools.

Cable and Gloves

The HO must always wear gloves when guiding cables, and positive pressure should always be maintained on the cable, otherwise it will cause a loop to form and result in cable fouling.

Hoist Hook

The HO should keep the cable and hook in sight at all times to prevent it from becoming entangled with objects in the hoist area, or hitting people in the eye.

If this page is a photocopy, it is not authorised!

Delivery

The hook or apparatus should be delivered to the ground or deck, and not placed in the hands of the ground party except in the most controlled conditions, or everything gets delayed and injuries on the ground may result.

Retrieval

With the Rescue Sling, the best way to bring people into the helicopter is backwards.

Securing

Once inside the aircraft, extra cable should be winched out to ensure that the person could be secured in a seat before the Strop is released from the hook.

Slack Cable

During hoisting from a vessel, particularly with high seas, cable may pile up on the deck as swells lift the ship. This slack must be kept to a minimum to stop the cable from snagging.

HOISTING FROM A VESSEL

Before you start, it is normal to overfly and confirm the vessel's identity (and punch its position into the GPS). The helicopter then needs to be set up at very low altitude into wind on an approach. The vessel may be asked to hold an into-wind heading, or maintain the hoisting heading (wind 30-45° off the port bow).

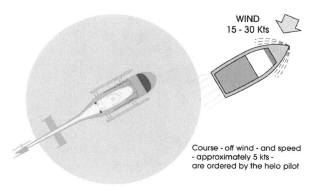

WIND 15 - 30 Kts

Course - off wind - and speed - approximately 5 kts - are ordered by the helo pilot

The concept is simple and involves a tear drop procedure adjusted for windspeed, using the GPS, timing and radar. If the aircraft is already positioned downwind and the vessel position is known, a straight-in approach can be flown from 2-4 nm out. Hoisting checks should be normally completed before this point.

Approaching The Vessel

Assess the wind speed and direction to determine:

- The outbound heading (reciprocal of the wind - 20°)

- The outbound timing

- The inbound heading

- If the vessel has been positively identified and distance to it can be accurately assessed with radar or GPS, fly a straight-in approach

- Brief for the approach (headings, speeds, altitudes and timings)

- Call for descent to 500', speed 80 knots

Outbound

- Maintain 80 kts and 500 feet until time is up or GPS reads appropriate distance

- Commence Rate One turn inbound

- Roll out on the inbound (into wind) track

Inbound And Final

- When level, descend to 200 feet, slow to 65 knots

- At 1½ miles (GPS or radar) descend to 100', slow to 50 knots

For Sikorsky S-76 speeds:

- 80 knots fly ....+1° attitude

- 65 knots ....+3° attitude

- 50 knots ....+4° attitude

- Hover ....+7° attitude

On Top

- Enter position into GPS and set read-out to guide you back to target.

- Start timing (to backup GPS)

- Turn to the outbound track

- Helicopter arrives on top of the ship, free of obstacles; lowering a weight with a guideline

- The line should be handled by a member of the ship's crew; helicopter goes to one side of the vessel

- Once in position, hoisting begins (min altitude 50 feet)

- A member of the helicopter's crew or equipment comes down; the ships crew continue to take in the slack

- When a member of the helicopter's crew or equipment is at deck height slowly pull until he or it is on board

After approaching the boat from astern (downwind) and entering the hover off the port side, aft of amidships. This allows the pilot and HO (on the right side of the aircraft) a full view of the boat during the evolution.

Lowering The Rescue Device

The rescue device will be lowered from the right side of the aircraft:

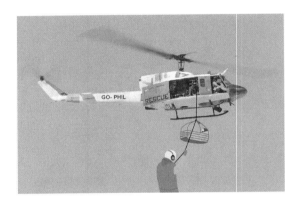

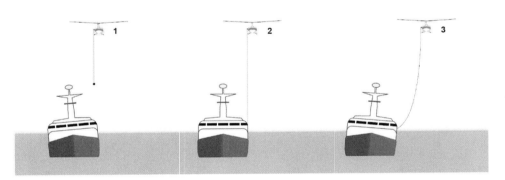

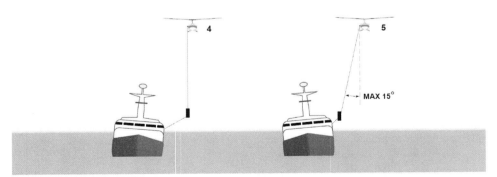

Helicopter arrives on top of ship, free of obstacles, lowering a weight with a guideline. The line should be handled by a member of the ships' crew; helicopter goes to one side of the vessel.

Once in position, hoisting begins (min. altitude 50 feet). A member of the helicopters' crew or equipment comes down; the ships' crew continue to take in the slack

A REAL RESCUE

This (from Paul Johnson) sums it all up very neatly:

> It's 11.00 am one day last week. We've just returned from the coffee shop at the terminal. All 5 of us were mesmerised by the most beautiful woman anyone's ever seen......wives and girlfriends excepted ;-))

> The co-pilot and I are in the pilot's room, reposing on the genuine fake leather sofas that replace the private bedrooms provided at all other company bases. He's rattling on about some aviation related issue, and I'm pretending to care and secretly daydreaming about the woman in the coffee shop…I just get to the part where she was saying…???? when the "bat phone" rings, and we haven't quite reached the… "I don't care" stage…

Aside: SAR duty days are divided into 3 periods.

- *Arrival…* pre-flight, with engineer after his daily inspection. Weather and NOTAMs, followed by one or more visits to the coffee shop, followed by checking the Internet for another job! (Doesn't matter if the one you have is OK…old habits die-hard! ;-)). This period lasts about 5-6 hours, and is followed by:

- *I don't care anymore*…this is when you're sick of waiting for an interesting mission to come up - now you can take it or leave it! This lasts about another 4-5 hours. Followed by….

- *They better not call now!* The machine is clean and tied down, all plugs are in, and any activity now means we'll be going into overtime. Every mission means an engine wash and winch cable wash at the very least, as it's a highly salt laden atmosphere

But we haven't reached *I don't care anymore* yet, and the next thing is, the WO runs by shouting "We have a mission" followed by another rear crew calling by cell phone to say, "We have a mission!" followed by the other RC running up the stairwell and saying, "We have a mission!"

Naturally, I walk out and shout: "Do we have a mission?"

Four voices shout in unison! "Yes"

I put my flight suit on and tie my boots. I'm still on the 2nd floor so I shout down "Mountain or sea?"

I get 2 "mountains", a "sea" and one "I don't know". I can see things are moving along smoothly, so I head for the helicopter. The engineer has pulled the plugs and tiedowns, and is standing by.

I climb in and we have our usual discussion about GPU or no GPU (it's a huge prehistoric thing, and the guy gets a bad back every time he drags it around. The airport authority won't let us leave it co-located with the ship).

So I call for start clearance, and get number 2 flashed up while P2 is finishing a last minute walk around with the engineer. Then I start number 1 while P2 is strapping in. I finish the flow pattern and P2 reads the checklist (We do a challenge and response as a check list, not a do list. I've already done the doing, this is just confirmation!)

We don't get co-ordinates until the last minute, if at all. Often it's "in the region of" such and such a town, mountain, beach, etc., and we fine tune it en route. In this case we got co-ordinates from RCC but they were wrong anyway. We only operate up to 12 nm out to sea on standard missions, with a standard fuel load, so PNR/CP issues are all standard figures. Outside this distance, we can take as much time as needed for FP purposes.

"Rescue 31 is ready on the South ramp. VFR to the West with Kilo"

"Rescue 31 is cleared via VFR 2, and then Westbound, not above 1000. From present position, cleared for takeoff"

We're around 8-9 minutes after the call. We launch from the ramp area, and hang a left at 300'. I have the a/c and Victor radio, while P2 gets all the info he can from RCC. I head in a generally Westward direction, actually heading South of West to avoid the mountain at 12 o'clock. The P2 finally comes up with a rough heading and throws it in the GPS. It's about 20 minutes away.

The only thing we have heard from the scene is that it's one or three victims, but no-one is sure, and he/she or them, are either half way down a cliff face, or have fallen onto the rocks where it levels out into the ocean.

We hit the co-ordinates and find nothing, so we cruise up the coast a little way and see a bunch of rescue vehicles parked on a high road, winding down the ravine. The on-scene commander calls on FM and warns us of power lines in the vicinity, so we do a high recon and see them running pretty close to the victim (one female) and then angle away. Behind the primary wires is a SWER line maybe 100 metres back up the ravine.

We do a second recon and decide it's too risky for a let down between the wires, so we'll enter the ravine from seawards, low and slow, following the ravine bottom upwards, under the wires, and set down on a flat spot about 2-300 feet below the scene. In the event of a no landing possible due rough terrain, an RC will hover dismount, and mountain goat it up to the scene, and take control, possibly looking for a place she can be stretchered to, away from the wires. As it happens, when we get to the flat spot, it's OK for landing, and large enough for a second machine if needed, so we park to one side and shut down. This is not going to happen real quick.

We have 2 other machines colocated at our base, plus another 2 within 60 nm in each direction, so a shutdown is approved as the lesser of two evils. If we keep turning and burning for an hour, we might have to pull out for fuel, and delay the rescue, so the minimal risk of our well-maintained machines not starting is justified.

The RC are all fit enough for Olympic tryout, and after 3 minutes, 2 of them are on scene (I'm talking of a nearly vertical rock face…I estimated about 30 minutes if I had to do it myself! ;-)) The rear crew need to be fit, and I mean *really* fit, because they can be called upon to run up a mountainside and pack a victim back down in a litter, as some mountain operations don't allow the use of a winch for crew safety reasons.

The RC on site explained that the wires are actually around 30 metres from the overhead position, and then angle away fairly quickly. The RCs agree it's tight, but do-able from their point of view (we have a basic rule, that if anyone in the crew says's no-go, it's a no-go. In such a case, we'd ask attending ground crew to go for option B, whatever that might be). As there was no viable Option B in this case, it was good to hear that the guys thought it was do-able.

We briefed the operation by radio. It took around 6 minutes, with everyone playing Devil's Advocate, then finding a solution for each eventuality. When everyone thought we had all the bases covered, we re-started with no problem.

We departed the LZ vertically, with the tail turned out from the hill to give the WO a clear view of the tail. He monitored the distance from the rear SWER line while the P2 and I focussed on the major wires ahead. Basically, we just climbed vertically up the hill and arrived at the standby position, more or less level with the major wire group, level with the road, and about 30-50 metres out of the vertical. When the WO was ready, he called "Move in". At this point, I still had visual contact with the victim/scene and could use it to maintain position. As we moved sideways, I started to lose visual contact and called "Contact lost!"

The WO then takes over as my eyes, and gave me the following calls:

> *Forward 2… Right 10….*

These figures are not feet or metres, but figures that represent distance, and rates of closure. The speed at which I close with his chosen position will determine the next call. It should only need a few calls to get me in position. If it´s gusting etc. It might take a few more. The key here is it's better to move too slowly, than too fast, particularly when you have live bodies on the hook. In the case in point, we have trained rescue personnel in position to receive the guideline. They've humped it up from the flat spot, to the ravine where the victim is lying. Often the first 'winch out' is a RC with no guideline, but that's more dangerous. If we can get a guideline into the hands of a responsible person, it's far safer.

In the following calls, imagine a pause between the numerical call-outs, long or short, depending on how quickly/accurately I'm responding to his instructions. Basically I'm a kind of second-rate autopilot for the WO, but lack the circuitry to perform exactly the same under all conditions ;-)

WO calls:

> *Forward 20…10…5…. Forward 5…. 4….*
> *3…Forward 3….2…. steady…forward 2…forward*
> *1…steady….(hold position)…right*
> *1…steady…back 2…back 1..Steady!*
>
> *Hold position!*

We're now in the vertical position, or the position the WO prefers for the next step. WO Calls:

Guide line (aux line) going down….Guide line ½ way down…Back 1…Steady!

Back 1…. Steady …Guideline on the ground (for static discharge)

RC has the guideline…

The WO continues his callouts. If we were in a 'normal' rescue situation, where we have no-one on the ground yet, we'd have to put the guys down like this:

One…(first rescue guy) leaving cabin… One's…on the skid…winching out….

One's ½ way down…One's approaching the ground…. Steady! …One's on the ground!

One's unhooked….Guide line is attached!

Move out!

During this procedure, the pilot is trying to maintain a perfect hover. Sometimes it's damned near perfect, other times it's a battle, but you cannot under any circumstance lose the plot. If you have an inadvertent gust, swing etc. you have to put it out of your mind and get immediately back in the game. As I move out to standby position, I call "contact" when the scene comes into view.

As we are moving out, the RC is feeding us more guideline, and we have a loop of it from the ground to the ship's winch hook. The RC we just put down will do his medical assessment and decide what he wants. If he has willing and capable hands available, he'll tell us to standby, using hand signals. If he needs another RC to help him, he'll signal his needs. In this case, we'd move back in and put another RC down, then move out again.

Mountain rescues can be more dangerous than sea rescues because, if a line breaks, or we suffer momentary control problems, it means the RC on the line might be in real trouble. The typical ravines we work in are real tiger country - rocks and cross-gullies everywhere, so we try to limit the guys going down via winch to an absolute minimum. The RC normally will try to accomplish things on his own if possible, and avoid the exposure to his colleagues.

OK, so here we are… We have crewmembers on the ground. We are attached to them via a guideline, and we just wait. We have the option of recovering the guideline if it's appropriate for the situation, and departing for a kind of traffic pattern/circuit, which is exactly what we do, if hover power required takes us into and beyond 5-minute power.

If we are in 5 min power, the co-pilot calls out the minutes starting with 3 to go, two to go, etc. I relay to the WO so he can plan his operation. In the incident in question, we hold position while the RC and helpers stabilize the victim. Then the RC signals for a litter.

Now, while we were at the 'flat spot' with the scene commander, I told the RC that, once we get the victim on board, we'll depart for the hospital and recover the RC later. They can either wait at the flat spot, or return with the ground pounders to the hospital. I asked them to call us after the operation was complete via cell phone, and tell us what they want to do (we have cell phones installed in the machines, hooked thru the comms system to your helmet). Again, what we don't want over this kind of terrain is to risk a RC if we don't have to.

So, the WO calls:

Move in

I call:

Moving in - Contact lost

He says:

Right 20…forward 3…right 10…forward 2…. stop forward…right 3….right 2…steady… Hold position!

For the litter/victim recovery, we don't have to go to the true vertical as we might have to do when we deploy the RC. The winch limit for recovery is 30° from the vertical, but we use a 20° maximum, to account for control error, so if we're within the 20° limit we're ready to go. He asks me if the hover is stable (remember we're on the crest of a canyon with 30-35 kt winds). I say *Yes*

He calls:

Winching in

and the litter is recovered with the RC on the ground controlling any swing with the guide line, and the WO still giving me 'steers':

> *Right 1..steady…back 1….steady…etc…*

WO calls:

> *Litter ½ way up…Litter approaching the skids….Litter at the door….Litter coming in…..Litter in cabin….*

Wait a short time and he calls:

> *Cabin and victim secure…Door closing….*

Now we can start to accelerate up to 60 kts in this particular machine. The WO then calls *Doors Secure!* We check the door lights on the CWP and if lights are out, we fly away to the hospital, securing winch supply power as the doors are closed. But in this particular rescue…We don´t fly away!

We back off the major power lines a few feet, and I start a left slide down the mountainside, back into the canyon floor and depart the way we came in. We get established in the climb and the co-pilot asked why we didn't just take off straight ahead, climbing over the power lines, as we already had some altitude. In fact, we went down, only to climb up again.

I was tired, so I said it was "secret Captain Shit!" Once we got going back to the hospital I asked him to fly for a while.

Explanation: of why we didn't climb out straight ahead, which I passed on to P2 after I got my energy back!

> Here we are, heavier now, as we recovered the litter with an 'economy size' victim. The torque is in the late seventies, early eighties and we can't back up, and get a run at it to get V_{TOSS} well before we hit the wire zone. These wires are multi rack HV lines about 80-100 ft from top to bottom…

> As we can't back up to get a run at them because of the SWER line behind us, it means both full power and a vertical climb (not a good scenario when facing wires) or taking off towards power lines only 10-15 meters away and hoping we don't lose an engine.

> It probably sounds incongruous to say that there's a risk of engine failure, and a possible wire catastrophe, considering we've just been hovering OGE in hostile terrain for the last 15 minutes, but this is the fine line in the wonderful world of

helicopters. We do what we have to do, when we have to do it, in the calculated risk sense, but we never take a single risk we don't have to…

In this operation, we're far better descending within the HV diagram, low power applied etc., passing the terrain we flew in over, than heading out climbing and accelerating over heavy duty wires which may or may not be exactly where they appear to be.

OFFSHORE

What people usually mean by offshore flying is the support of oil rigs out in the middle of nowhere by shuffling people and equipment to, from and between them, often in seriously bad weather.

However, although you might think an oil rig is an oil rig, there are many variations on the theme, ranging from a simple wellhead (below):

to a large conglomeration of living quarters:

The wellhead, being so close to the water, means that that will likely be where you end up if an engine fails on takeoff unless you are truly in a Category A machine, not that the other one is much better, even if it is 120 feet up from the surface.

How the operation is run can vary - in some companies, you go out on to the rig for a month and hop around the field, and in others you go out and back from day to day. This will naturally depend on how far the oilfields are away from shore, but the end result is that there are regular shuttles at specific times. There is a growing trend, on the North Sea, at least, for longer flights up to 300 nm from the point of despatch and sometimes taking over 6 hours to complete. If the weather is bad and the swells heavy, there will also be some unscheduled field moves (*shuttle operations*) because the boats that usually do the job will not be able to move. Indeed, there are many similarities between offshore work and supporting the oil patch in the North of Canada - the only difference is that, instead of tundra, you have water underneath you, and you're still not guaranteed a landing spot!

IFR is normal procedure in many places - by day, the main hazards are a low cloud base and obstacles, although these are rarely over 500 ft. The weather limits are generally accepted to be 200 ft and 0.75 nm, because you need to be able to miss the target (and other rigs) if you have to go around. By night, you could lose contact with the surface by getting fixated on the rig, or get inadvertent IMC in the final stages of the approach. Radar is often used as an aid to navigation and approaches, although it is not certificated for either.

The landing sites are elevated, of limited size, surrounded by obstacles, so the approach is a decelerating manoeuvre terminating with zero groundspeed. As a result, approach angles are steeper than for onshore approaches.

HELIDECKS

Oil rig construction is governed by ICAO Annex 14, Volume 2. Naturally, some decks were made before that document came into force, so your company should have a *Helideck Limitations List* (HLL) with all the non-compliances that apply to your operation (under Part OPS, in Part C of the Operations Manual).

A *remote installation* is over 40 miles from the nearest manned installation or airport/heliport. A *Normally Unattended Installation* means what it says, and the number of passengers should normally be restricted to those being delivered to the NUI, or being collected from it, although there should always be a minimum, enough to handle the helicopter on the deck (see *Personnel*, below).

Installations all have one thing in common, though - restricted helipads potentially surrounded by obstructions, and covered in birdsh*t or water, depending on your location in the world (some may have a net stretched across to provide increased surface friction or give people who fall over something to hang on to). They also have two stairwells - the primary is for normal use, and the secondary one is for emergency access. In many ways, the process of landing on them can be compared to landing on mountain pads, leaving aside the obvious differences. You are going for the only good landing spot within miles, and if you are operating somewhere hot there are also some power restrictions.

As a result, offshore decks should not normally be used for these emergencies:

- One engine inoperative
- Partial or restricted power
- Flight control malfunction
- Fire

If you need to make such an emergency landing, especially with a power loss or control difficulty, you should try for the best lit suitable area available at a manned airfield, where engineering support can get to you. For something like a hydraulic system failure, provided you can still fly the machine (say with a #1 hydraulic system failure on the B212), you could land if nothing else is on the deck and other machines stay away until you have shut down (if you have ever tried to hover a machine with non-working hydraulics, you will know why).

To stop you landing on the wrong helideck (especially when they are near international borders), the last item on the Landing Checks should be "Landing Site Confirmed". In other words, a *landing ban* is in force until the non-handling pilot (or *Pilot Monitoring*) has agreed that the helideck has been correctly identified. There will be markings on the helideck or on the sides of the structure.

Obviously, you can't inspect every helideck before you use it, so Company authorisation to use one is usually allowed if the inspections are done by accepted people or organisations, such as the BHAB or HAI. You could also use a Helideck inspection report conducted by a trained Company officer, another operator or an approved organisation, such as an Oil/Gas operating company, if it has been done as per the minima in the Ops Manual.

Where there is incomplete information, a limited authorisation based on what is available may be issued by the Company before to the first helicopter visit. Naturally, full authorisation should follow as soon as possible.

PERSONNEL

Important people on a rig include the *Helicopter Landing Officer* (HLO), who controls all movement on the helideck (including giving landing and takeoff clearances) and the guys in the Radio Room, who tell you what the weather and payloads are like.

Between the personnel around, these functions should be covered:

- An offshore Installation Manager (OIM)
- A Helicopter Landing Officer (HLO)
- A person holding a Restricted Aeronautical VHF Licence
- Two persons who have attended the Basic Firefighting course
- Two Offshore First Aiders

Onboard a helicopter going to an NUI, the HLO should be seated so as to be the first to disembark. After doing so and ensuring that everything is safe, the HLO will give clearance for the remaining passengers to disembark. The reverse applies before departure, when the HLO is the last person to embark.

PROCEDURES

The following general procedures apply. Landings may be made under all conditions (including on tow) except when a rig is jacking up or down and or perforating.

- When a rig is **perforating** (and using explosives), all radios and telephones should normally be shut down. Helicopters should maintain strict radio silence with no transmissions within one mile of the rig. Recheck that the radar and transponder are on Standby

- Landing on a drilling rig when **burning off** is in progress is acceptable if clearance is obtained. If the prevailing conditions are not suitable for landing the flame can be shut down on request

- Due to size restrictions and the difficulty in seeing the guy wires, helicopters should only land on a well head while **wirelining** is in progress in dire emergencies. *Extreme care must be taken*

- Due to local magnetic fields, **compasses** may be affected to varying degrees. The magnetic compass be OK once clear of the platform, but the gyro compass may take some time to re-synchronise, so use the compass control box, if you have one, to reset it again (the expanded rose display on the GPS is also recommended). Re-synchronising the gyro compass should be done as soon as possible after takeoff, especially on night flights and in reduced visibility. Another option is to unslave the gyro just before landing, but be sure to reslave after take off. A further effect is that the VOR course bar may align itself with the compass card and not the aircraft heading, which can be misleading when the compass is not synchronised

- A **retractable undercarriage** is normally left locked down for any legs less than 6 minutes, because the increased wear and tear makes it uneconomic unless the sector length justifies the reduction of drag/increased cruise speed. In other words, raising the undercarriage during shuttle operations (i.e. field moves) is the exception rather than the rule

- People must be **clear of the helideck** before landing or takeoff, and should not move onto it until a pilot gives a Thumbs Up signal (in many companies, this is always the PIC). The last person to move off before a takeoff must give a Thumbs Up to the pilot to indicate that the deck is clear and that no abnormal situation exists (HLOs with radios may substitute a radio clearance for the Thumbs Up). Clearance to land is *always* required

- Whenever a helicopter is landed or parked on an offshore helideck or platform, the **tail rotor must be positioned** *well clear* of the secondary stairwell, so if passengers have to rush for it, there will be minimal chance of them running into the tail rotor. It must NOT be assumed that people entering the deck will approach only from the upwind/nose access way. *The secondary stairwell is for emergency use.*

- Normally, with the exception of helidecks, overflying offshore platforms, drilling rigs, barges, tankers and flare and vent stacks (whether or not in use) should be avoided unless weather or operational reasons *require* you to fly over them.

- Before and during landing and takeoff, and at all times whilst the rotor is turning, when a helicopter is on a helideck, cranes must be stationary, with the boom stowed in the cradle, or boomed away from and clear of approach and takeoff paths and the helideck, with the operator standing clear of the controls and in view of the pilot.

Takeoff & Landing

In a two crew operation during daytime, the pilot with the best view of the surrounding or adjoining structure or obstructions should do the landing or takeoff at the discretion of the Captain. At night, the one best situated to make the landing should *always* do so unless the Captain decides that other considerations make it safer not to (as with an inexperienced co-pilot in low visibility conditions).

TAKEOFF

The helicopter should be positioned as close to the deck edge as possible without losing visual reference, ideally with the rotor tips at the forward edge, so there is at least a little space to land back on to if you have to. However, some pilots get as close to the edge as possible, with the rotors over the water, and the toes of the skids just inside the edge of the deck, to catch the wind that is deflected up into the rotor, in theory providing some additional lift. However, it is debatable as to whether this covers for the loss of ground effect. Probably not, below 10 kts.

When heavy, with little or no wind, and it's very hot, get to the edge, and *move nothing* until absolutely necessary (and even then, be very smooth). Eventually some lift will come along - the helicopter will let you know. Smoothly lower the nose to move off and do nothing quickly. If the torque available provides enough power, the takeoff should be made in one continuous movement. Like slinging, where you know whether a load is a good lift or not, you get to know when the old girl can do it. However, it is more prudent to establish a maximum torque limit of 85% IGE to ensure you have enough power in hand to get to 15-20 feet before rotating (this is policy in some companies).

To avoid flying over obstructions below and beyond the deck edge, the final takeoff heading and position will be a compromise between path required for least obstructions, relative wind, turbulence and lateral marker cues.

The helicopter should continuously move vertically from the hover to the Rotation Point (RP) into forward flight. However, if this is too dynamic, you risk losing your visual cues during a rejected takeoff, particularly at night. On the other hand, if it is too slow, there is an increased risk of contacting the deck edge if an engine fails at or just after RP, but this is mitigated somewhat with height (hitting the deck during an engine failure is something everybody is paranoid about). The climb to RP is best made with hover power plus 10-20%, which is why some companies have a maximum of 85% (the JAR OPS requirement of a climbout capability at 1000 feet means that vertical acceleration is possible and you won't need to exceed any limits). This both assists with edge clearance after an engine failure at RP and minimises ballooning if an engine fails after it. However, it may vary for individual types.

Whenever possible, takeoff should be made into wind - 10 kts of headwind can allow an extra 5 ft edge clearance over zero wind (and a delay of one second in pilot reaction time can lose up to 15 ft).

Once committed to the takeoff (usually on rotation), you must, if an engine fails, rotate the aircraft to the optimum attitude for the best deck edge clearance (check the Flight Manual). The optimum RP (which varies between types) ensures that the takeoff path will continue upwards and away from the deck with All Engines Operating (AEO), while minimising the possibility of hitting the edge if an engine fails at or just after RP. Again, a low RP means you might hit the deck if an engine fails, and a high one means you might lose any visual cues. If the helicopter has a low mass, it may balloon to a significant height before the reject action has any effect, so you should select a lateral visual marker and maintain it until the RP is achieved.

Many pilots use a miminal climb rate and accelerate as quickly as possible to flyaway speed (typically 60 kts, which they consider to be the CDP). Over unlit unknown terrain, use an instrument takeoff profile:

- Attitude Indicator set to 5° nose up on the deck

- A vertical climb above the takeoff area maintaining visual reference and with at least maximum continuous power set, until a positive rate of climb is shown by pressure altimeter movement

- Simultaneously, a gentle but positive attitude change to hold 5° nose down allowing the aircraft to accelerate to 55 kts (depends on type)

- At 55 kts, change to 5° nose up, to climb at the best climb speed (V_Y). Climb straight ahead until at least 500 ft above the surface

LANDING

The approach should be started into wind to a point *outboard* of the helideck, with the rotor tips kept clear until you approach this position at the requisite height (type dependent) with around 10 kts of groundspeed and a minimal rate of descent (around 250 ft/min is good). The aircraft then passes over the deck edge and into a hover over the safe landing area. This way, if the engine fails before LDP, you can carry out a smooth nose over, and set your power for the flyaway. If it quits after LDP, you already have a level aircraft, power set and no need to be nosing over or flaring. A steep(ish) approach, like in the pictures, will put you in the most favourable position for a landing with the least speed.

If this is impracticable due to obstructions or wind, the *offset procedure* involves flying to a hover about 90° offset from the landing point, at the appropriate height and maintaining rotor tip clearance from the edge. Then fly slowly but positively sideways and down to a low hover over the landing point. Normally, the LDP will be when you transition over the edge of the helideck. In fact, as with mountain flying, there are many schools of thought about approaches. I favour the steep approach, with the H just forward of the end of the pitot tube.

The gate is 500 feet (plus the deck height), half a mile and 60 kts (plus the wind speed). At night, start the descent when the lights split into a rugby ball shape. Although you might have to start down at around 500 fpm to keep the closure correct, if you pull in power early and aim to cross the deck at around 250 fpm, there is very little collective movement at the end, and no flare, so you won't hit the tail on the deck when you get into low hover, which you need sometimes when it's really hot.

Coming in steep(ish) means you have some momentum behind you (from gravity) if you have to make a dive for it, and you are not using as much collective as you would with a shallower approach, with which, if an engine stops, you are going to go nowhere but down - your collective is already high, and dumping it takes you down quicker. This is more important if the deck is only a few feet off the water. "Too Steep" is not an angle - to me it's where you don't have enough speed to fly away if you need to.

This is from another pilot, paraphrased from the web:

> "The less you have to do in terms of movement of the flying controls, the less needs to be done and the more control you have of the aircraft, and the better chance you have. The fast and shallow mindset, with large collective/cyclic movements, will put you in trouble."

> *My first priority on an approach is to fly it in such a manner as to preclude hitting something or losing control due to uncommanded yaw. This means no fast approaches, and only as shallow as absolutely necessary to see enough of the rig to fly an accurate approach. Wind speed and particularly direction are important factors to consider when planning the approach. The second priority is to avoid overtorques/temps/speeds, which can be done (again) by playing to the wind and taking things slow. After these considerations are satisfied, I will choose (if available) an approach path/profile combination that will allow in the case of a single engine failure, a nose-over to V_{TOSS} and climb away (in the early stages of the approach) or (from short final, probably a hard) landing to the deck whether or not with an overtorqued transmission and/or burnt engine. In some cases (heavy load, no wind) this is not continuously possible, but the probability of losing one engine during such a short segment of the approach is far smaller than that of hitting something you can't see, and so receives less consideration.*

> *The end result may be a straight-in approach, an approach to a spot beside and above the deck followed by a slide over the deck, or an approach terminating in a curved segment of about 30°. Often I fly the last 100-200 feet out of trim for better visibility – no problem if you keep in mind and anticipate pedal requirements."*

And another:

> *"I aim for the far edge of the deck, and at the angle which I use for the approach, that puts me near the center of the deck at a hover, which is right where I want to be. If I fall from there, I'll stay on the deck. If I aim for the near edge, I'll end up over the water, and I don't want to be there. I sometimes fly with pilots who want to approach fast and shallow, and that scares me. If you don't get slowed down exactly right, it can get really, really ugly on the bottom. A slow approach at about 250 fpm descent lets me come to a hover with very little extra collective needed, no hairy flare to stop, and a constant heartrate."*

The helideck can often be sloped away from the centre, to help with drainage. The slope is not severe, but if you land with your nose up the hill, so to speak, you run the risk of hitting the tail rotor against the perimeter fences. Landing "downhill" will artificially raise the tailboom away.

Tip: Try to land inside the circle - that means they only need to paint that area!

Just to make life difficult, depending on the wind direction, you may have to fight with the exhaust from the turbine engines that keep the place running as it flows across the helideck. The turbines are often the industrial version of the Avon, as used in the Lightning, so they are pretty powerful.

SHIP'S DECK LANDINGS

This also covers floating helicopter decks.

Caution: Extreme caution must be exercised!

First of all, you have to find the ship! Unfortunately, once they've got full steam ahead, ship's masters are very reluctant to slow down for anybody, so be prepared to look for your target a fair bit away from where you (or the owners) think it should be (shipping agents lie like duty-free price lists). However, they may turn into wind a little bit, if you ask them nicely and remind them that you have their relief on board. Aside from knowing where the deck is, you will need a radio frequency to talk on, either the customer's own or a general marine channel (16). This is not only so you can ask where the heck they are, but so they can give you the wind and how much the deck is going up or down (that is, pitching and heaving, for the technical). *Do not land without permission.*

Flying around vessels is much like doing it round mountains or buildings, especially the latter, which are more slab-sided. The wind does strange things round them all, so have the normal escape route handy, by approaching obliquely. If you approach from the bow, with the vessel moving toward you, the approach can get very steep very quickly, and approaching from the stern means it can get very shallow, as it moves underneath you. There will be turbulence in the lee (and elsewhere), so approach to the windward side, getting into the hover before you position for landing as the highest point of the swell is reached (not when the boat is rising) - use your peripheral vision to keep an eye on the deck, and focus on the horizon, as normal. Look out for ground resonance, and a slippery deck, which will likely be sloping as well. Stabilisers will compensate for deck movement, so turn them off once you are safely down.

Landing Areas

The landing area may be a dedicated flight deck at the stern, or more commonly a clear area amidships, marked with an H in a circle or other distance rings. As a general rule, they are very, very tight (ship size has nothing to do with it). Side rails can be lowered, so request it early enough for more tail rotor clearance.

Decks of tankers usually slope downwards and outwards from the ship centre line, plus a camber of increasing proportion towards the deck edge. A helideck on a small tanker may take an external load, but not necessarily a helicopter.

Drilling barges usually have unobstructed decks, whereas derrick and pipe laying barges tend to have aerial installations nearby, in addition to fixed and mobile cranes. Tugs and support boats are often moored alongside barges, usually on the downwind side, and can be a significant hazard to helicopters, especially if you have to take off over them.

Ship Motion

Vessels underway have 3-dimensional motion, made up of *mean line of advance, pitch, roll* and *heave*, the latter three being most important. Pitch and roll are the same as in aeronautics, while heave is vertical motion, expressed in feet. A loaded supertanker hardly moves up and down at all, but an empty one may roll if the wind is abeam.

In heavy seas, the pitch or roll, or a combination, can severely limit helicopter operations. You should try to get actual readings from the vessel and watch its movement for at least two minutes before landing. Generally, roll and/or pitch over 7° (3.5 each way), and heave in excess of 10 feet, especially a combination of unpredictable pitch and roll, will preclude a landing - do *not* land if the pitch and roll exceed your sloping ground limits. Heave over 15 feet is definitely dangerous and can lead to high touch down forces. Most maritime helicopters have wheels and oleos to absorb them.

When the deck is moving, the optimum takeoff movement is when the helideck is level and at its highest point, e.g. horizontal, on top of the swell. Collective pitch should be applied positively and sufficiently to make an immediate transition to climbing forward flight. Because of the lack of a hover, the takeoff profile should be planned and briefed before lift off.

RELATIVE WIND

The combination of ship course and speed, and the actual wind, gives rise to an airflow over the vessel called the *relative wind*. Mariners usually express it in knots and its direction from the ships bow, in degrees. Whether the wind is left or right of the bow is detailed by a colour: red left, green right (as for aircraft position lights). Example: Relative wind is 15 knots, red 45. Decode: Wind is 15 knots, 45° left of the bow. Obvious telltales are the ship's flags and the exhaust plume.

Note: Watch out for the ship turning while you're on the deck. The crosswind you landed with can become a tailwind very quickly.

Communications

Initially by radio, on Channel 16 (Maritime Distress, Safety and Calling channel), then organise another channel for the operation. If voice communications are not established, look for the following visual signals:

- Flag "Hotel" which is red and white, flown at half mast. It means "I am preparing for flight operations"

- Flag hotel hauled close up means "I am ready for flight operations - proceed"

Shapes may be shown that indicate to other vessels the restricted manoeuvring ability due to flight operations. These are a ball, then a diamond, then a ball hung vertically from a yardarm (where they hang pirates from).

Approach & Landing

Check out the landing point. Establish the motion of the deck in terms of pitch roll and heave, then plan the approach into the relative wind if possible, and clear of the vessel for as long as possible. Land with caution, using sloping ground technique. Once landed, do not follow the horizon with the rotor disc. Centralise the cyclic and keep it there. Do *not* decelerate to ground idle - the helicopter may slide on the rounded deck, and having flight RRPM to get you out of trouble is most important.

Departure

Align yourself with the way the vessel is going (even if there is a crosswind), and avoid the superstructures on your way out.

NOTES

OPERATIONAL STUFF

This chapter contains some hints and tips that should help you with your day-to-day flying. Some of it is required reading for Canadian commercial pilot exams.

VFR EN-ROUTE MINIMA

A US Army study indicates that it takes about 5 seconds to perceive a problem, make a decision and start a correction. At 80 kts TAS, therefore, you will move 676 feet and get still closer as you turn away from an obstacle, the distance being equal to the radius of the turn which, in this case, would be 984 feet, assuming 30° of bank, giving you a total distance of 1660 feet to cope with, which is not good if your visibility is only 1000 feet! So, in bad weather, *slow down*, certainly enough to give you at least a two-minute visual reference ahead, but in any case, never faster than what you can see in front. Approach a ridge at an angle so you have an escape route, and be aware of the slope.

Sunglasses with high contrast are a good tool for seeing cloud textures in potential whiteout conditions, as is something on the windscreen to help the rain slide off.

Flying VMC on top of cloud in single-engined machines is not allowed because performance rules require that, if an engine fails, you must remain in sight of the surface, and do a safe forced landing.

DO NOT LOSE SIGHT OF THE GROUND AT ALL IN THE ARCTIC.

In Winter you will have to deal with ice crystals, that will reduce the visibility to a mile or two, and you will have almost zero contrast on the ground. You have to be extremely careful, and watch out for the flat light conditions known as sector whiteout. From November to February, the days are VERY short, in contrast to the Summer, where the days are long... very very long.

Over Water

When crossing estuaries, make sure you see the other side before leaving the side you're on, so you get as little of the goldfish-bowl effect as possible. Also, keep sight of the shoreline. Out of sight of land, the visibility should be greater than 1500m by day and 5 km by night.

In the Canadian Arctic, you may be asked to cross vast areas of ocean, such as the Hudson strait between Quebec/Labrador, and Baffin Island, in a Bell 206L with pop-out floats. You will be nowhere near the gliding distance, and you will be very heavy. This will be in Spring and Fall and you will be fighting fog and low cloud banks, trying not to hit ships and icebergs.

LANDING SITES

In the UK, aircraft below 2730 kg MAUW don't need a licensed aerodrome, if the flights do not begin and end at the same place, take place at night, are not for training or are not regular services. Otherwise, licensed aerodromes must be used for scheduled services and training. Also, in UK, the owner of a piece of land does not need special permission to use it as a Helicopter Landing Site if certain conditions are met. It must not be in a congested area, or you will come up against the regulations. It must also be only for private or business use, that of any employees, or people specifically visiting for social or business purposes. Finally, no structure must be erected in connection with its use for helicopters, aside from temporary ones (such as windsocks), otherwise the Planning Permission (Zoning) people will become interested - there's no need to notify them of anything unless the land is to be used as a helipad on more than 28 days in any year. In fact, current planning regulations allow a helicopter to be used for personal, business and leisure uses "as many people use a private car" from the owner's dwelling house without limitation, making it exempt from planning control, provided the use is incidental, or ancillary to, the principal use of the land. Also, the local police should be informed, as well as the other emergency services, especially where the public would normally have a right-of-way. However, making more than four movements at a place in a relatively short time (a movement is a take-off or a landing) makes it a 'feeder site', and subject to strict standards-relevant if you're performing shuttle flights at a special event, such as Epsom races, or the Grand Prix. Pleasure Flying and Feeder Sites are considered in Chapter 3.

In the USA, and probably Canada, you can land a helicopter anywhere that has not been declared as illegal, with the usual provisos about low flying and reckless operation, but be aware that local restrictions may well override any from the FAA or Transport Canada.

Sites should allow you to make emergency landings without danger to persons or property on the surface, or significant risk to the helicopter and its occupants.

The following criteria should apply to all unlicensed sites, which are technical requirements that do not necessarily allow for low flying rules. A congested area is "substantially used for recreational and residential purposes", etc., which officially makes a golf course one, though you would be forgiven for thinking otherwise. A rule of thumb is 60% buildings and trees, but specifics haven't been tested in court yet, at least not in the UK.

There should be at least one approach and departure lane containing either no or only isolated obstacles, preferably not downwind, but see below. The lanes and landing areas should be big enough to ensure you can land, take off and reach a safe height so you can touch down into wind following an engine failure, while avoiding obstacles by a safe margin.

Try not to have marshland underneath the lanes because, while it may be soft, skids or wheels may sink in during an emergency landing, which is the last place you want dynamic rollover. In other words, the ground beneath the lanes must be suitable for emergency landings with respect to slope, softness, frangible obstructions, etc. Water is OK, provided the performance group is suitable or you've got the usual lifejackets, floats, etc.

The landing pad should be level, drained, with a grass or solid surface that does not blow up dust at (you should be able to drive the average car over it). Its diameter should be at least twice the length of the largest helicopter to use it, including rotors, as you will need to turn round your tail. Watch out for anything that may snag the skids, particularly on takeoff. Some people like the touchdown area marked with an H, but if the grid reference is accurate enough, it shouldn't be too hard to find it.

Finally, a couple of points to watch if you're ever tempted to land across or near railway lines, as you might if they're the only firm place around. The first is that your skids more often than not will complete an electrical connection used for signalling, and you may cause some confusion in the local signal box. The other is that trains use the track outside published schedules, so don't be surprised to see a humungous diesel bearing down on you unexpectedly.

FUEL

Jet and piston fuels mix differently with contaminants (particularly water), which is due to variations in their specific gravities and temperature. The s.g. of water, for example, is so close to jet fuel that it can take up to 4 hours for it to settle out, whereas the same process may take as little as half an hour with Avgas. As a result, there is always water suspended in jet fuel, which must be kept within strict limits, hence two filtration stages, for solids and water. The latter doesn't burn, of course, and can freeze, but it's the fungi that gather round the interface between it and the fuel that is the real problem - it turns into a dark-coloured slime which clings to tank walls and supporting structures, which not only alters the fuel chemically but will block filters as well. Not much water is required for this - trace elements are enough, although, in reduced temperatures, dissolved water will escape as free water, and look like fog. Aviation fuel is "clean" if a one-quart sample is clear of sediment when viewed through a clean, dry, clear glass container, and looks clear and bright.

Note: It has been found that when visible water is present in jet fuel containing anti-icing additive, the additive will separate from the fuel and be attracted to the water. After a certain amount, thought to be about 15%, the density of the new liquid changes so much that it is not identified as water, and will therefore pass through water filters, and not be detected by water finding paste. Where the ratio becomes 50%, as much as 10% of whatever is going through the filter could actually be water, which is very likely to get to the engine, since the filters on the airframe itself are not as restrictive.

Drums are also flushed with a solution containing sodium that passes through filters.

While on the subject of filters, they work on a certain pressure, which may be less sometimes, if you use a manual pump, or more if you use an electrical one. In other words, if the pressure is not as designed, filters will not do their job properly.

Fuel is actually a combination of various (very toxic) substances - pound for pound, it's more explosive than dynamite. Jet A, standard for commercial and general aviation, is narrow-cut kerosene, usually with no additives

If this page is a photocopy, it is not authorised!

apart from anti-icing chemicals. Jet A1 has a different freezing point and possibly something for dissipating static, used for long haul flights where the temperature gets very low. Jet B is a wide-cut kerosene containing naphtha, so is lighter and has a very low flash point (it's actually 2/3 diesel and 1/3 naphtha, but in emergency you can swap the naphtha for avgas to get pretty much the same thing). It contains static dissipators and is widely used in Canada. Try not to mix Jet A and Jet B - the mixture can ignite through static in the right proportions, as Air Canada found when they lost a DC-8 on the ramp in the 70s. The static can come simply from the movement of fuel through the lines. Jet A weighs about 5% more per litre than Jet B, but it gives you a longer range, as turbines work on the weight of the fuel they burn, not the quantity.

So, if you load the same amount of fuel, your machine will weigh more with Jet A, but if you fill the tanks, you will use fewer litres and less money.

JP4 is like Jet B but with a corrosion inhibitor and anti-icing additives. It was the main military fuel but is being superseded by JP8, at least in the USA. JP5 has a higher flashpoint than JP4, and was designed for US navy ships (similar to Jet A). JP8 is like Jet A1, but has a full set of additives (or e-numbers, if you like).

Aircraft parked overnight should ideally have tanks completely filled to stop condensation, but this is impractical if you expect a full load the next morning and don't have room for full fuel as well, in which case be prepared to do extensive sampling from the tanks. Half filled drums left overnight should not be used for the same reasons, but, in remote places (like the Arctic), fuel is a precious commodity and you think more than twice before discarding any (as it happens, drums are scarce too, and they may get used for all sorts of things, particularly diesel for drills, so beware - always smell the contents first). Full drums are usually delivered to a remote cache by Twin Otter or something, and they should be sealed straight from the refuellers - as you tend to use any remainder in a very short time, this can be minimised somewhat. Look for a fill date, as fuel over two years old should be looked at sceptically. Also look for a large X, which is the accepted symbol for contamination, although not everyone has a black marker with them.

Tip: If you are using a fuel cache that has been dropped by someone else, locate it *before* you start work - not when you need it, as it may be miles away from where they said!

Speaking of fuel caches, this is from Harald Sydness:

"On a job near Geraldton, in Ontario, I was flying timber cruisers around the bush. I had the machine literally parked outside my hotel room, with a small fuel cache. My crews were always four teams of two, so I'd fly out group A and B, then C and D, and in the afternoon/early evening I'd pick them up in the same order, always with coordinates and air photos of possible drop off and landing sites. The customer had determined that these were good enough - I'm sure there is a good story in there about that as well, but that is for later.

The job was said to last about 3 weeks, barring poor weather. Just me, no engineer, as I generally flew only 4 to 6 hours a day, it would take some time to get to the magic 100 in the Jet Ranger. When I was given the mission from the owner of the company, he gave me the map with coordinates and locations for the fuel caches, and I took that as gospel. Bad mistake. The first few days were normal, you get to know the customer, everything started going smoothly, you do your runs, fuel stops etc, and all is good, right up until the last week on the job. The distances are getting farther and farther, and now you HAVE to fuel up on the way back with every crew. As it was, I'd throw in as much as I possibly could when leaving the hotel in the afternoon, so that I would have to take only a little bit to make it back.

Well, on this particular day Murphy did strike, and almost as hard as he could. I did NOT follow common rules, and flew out, picked my teams out of the bush, and as luck would have it, they were the last teams of the day. We headed for the fuel cache, and tada! No fuel cache at all. I searched for about 15 minutes, doing mental math trying to figure if I could hit a different cache, but alas no. But I did figure I would have about 7 gallons of fuel left in the machine when I landed at the hotel if I bee-lined it there and didn't hit any wind. You get more and more sweaty and the chatting in the machine gets quieter and quieter as you get closer and closer to E on the fuel gauge, but my math did hold out, and we landed with the absolute minimum required by law.

SO, a tad ticked off, I call the boss, who in turn tells me that the fuel caches are verified by the supplier. Well, I say, call the supplier and ask, because the most important one was not there. Early the next morning we get the story from the fuel company, their driver deemed the one bridge he had to cross unsafe, so he put the fuel in a rock quarry about 5 miles short of the original position, and hid it with branches so no one would steal it, and forgot or neglected to tell anyone where it was, or indeed that it was in a different position than planned. From that day on, the job starts AFTER the fuel caches are verified, so that we won't have to buy new underwear again...."

Tip: An unofficial, but excellent (if not better) substitute for water paste or detectors with jet fuel is food colouring, which you can at least get in the local grocery store, even if you're in Baffin Island. All you need is one drop - if there is no water present, it will stay as a few lumps. If there is water, the colouring will go directly to the water droplets, and assume their shape, which are more visible anyway, from the colour.

Fuel Management

Very few aircraft will actually take a full load of passengers and fuel, so you need to know how long it will take between two points, find out how much fuel it will take, *then* fit the passengers in. *Do not put the passengers in first and fit the fuel in afterwards!* Not unless you plan to stop en route, at least. Of all the things there is absolutely no excuse for in Aviation, running out of fuel in flight is one of them! If you have to take less fuel, then you will have to stop and pick up some more on the way, or leave someone behind. If you take the same fuel anyway, you will be overweight, with not enough power in the engines to get you out of trouble, and *invalid insurance*.

Fuel flow will have to be adjusted if you plan to use specialised equipment in flight, such as heaters, or not use anything essential, like an engine.

Helicopters don't need aerodromes, and minimum figures reflect this. However, they are calculated for *level* aircraft. Odd attitudes, say when slinging, may cause a fuel boost pump to become uncovered and give you a nasty surprise just when you don't want it. On a 206, the unuseable fuel after a boost pump failure can be up to 10 US gals, which is uncomfortably close to the minimums. In short, under VFR, you should be *landing* with 20 minutes' fuel on board at normal cruise. For IFR and night VFR, it's 30 minutes.

For piston-engined helicopters, where the engine RPM is more or less constant, the Specific Fuel Consumption won't change much, but turbine engines change RPM constantly to maintain RRPM in varying conditions, so the best SFC, and range, is found at higher power settings, so the range speed in a turbine is higher.

EUROPE

Part OPS 3.255 requires your Company to have a fuel policy for flight planning and in-flight replanning, to ensure that every flight carries enough fuel for the planned operation, plus reserves to cover any deviations. Such planning must only be based upon procedures and data in or derived from the Ops Manual or current helicopter-specific data (e.g. the Flight Manual), and the conditions under which the flight is to be conducted.

Part-OPS 3.350 places a requirement on the commander to carry the planned amount of fuel - you may not commence a flight unless you are satisfied that the helicopter carries at least the planned amount of fuel and oil to complete the flight safely, taking into account the expected operating conditions.

Part-OPS 3.375 requires management of that fuel, including declaring an emergency when you have less than final reserve fuel on board (you can still use the fuel, but you must declare an emergency). *Fuel management* means you must check and record the contents regularly in flight, to ensure that:

- Actual consumption compares with planned consumption.

- The remaining fuel is enough to complete the flight - it must always be more than that needed to go to a heliport where a safe landing can be made, with final reserves.

- Check the expected fuel remaining on arrival at the destination. If it is less than required alternate fuel plus final reserve fuel, you must divert or replan the flight unless you consider it safer to continue to the (onshore) destination, where two suitable, separate touchdown and lift-off areas must be available and the weather is above the planning minima. In such circumstances, you may use the alternate fuel before landing.

You must declare an emergency when usable fuel on board is less than final reserve fuel. This does not mean that you cannot use it, but the rules are there to ensure that fuel management is carried out, as mentioned above. However, the intent is clear - in general, you *should* always land with at least final reserve fuel (the word *shall* may be used in your Ops Manual), and preferably more - proper planning should ensure that you don't get down that far, but sh*t happens sometimes. Declaring an emergency may allow ATC to bring you in with the reserves intact, by providing a more direct routing.

Pre-flight calculation of usable fuel must include:

- **Taxy Fuel**, at least what you expect to use before takeoff, including start, taxi and run-up, allowing for local conditions, APUs, icing equipment, etc.

- **Trip Fuel**. That required for the trip as planned. The difference between the helicopter's mass at takeoff and landing must be at least equal to this:

  - Takeoff and climb from heliport elevation to initial cruising level or altitude.

  - From top of climb to top of descent (e.g. the cruise), including steps.

  - From top of descent to where the approach is initiated, for the expected arrival procedure.

  - Approach and landing at the destination.

- **Reserve Fuel**, which must be on board at takeoff, but is not necessarily used. It consists of:

  - **Contingency Fuel** (unless otherwise approved):

    - For **IFR**, or VFR in a hostile environment, 10% of planned trip fuel.

    - For **VFR** in non-hostile environments, 5% of planned trip fuel..

  - **Alternate Fuel**, if a destination alternate is required (the departure aerodrome can be the destination alternate). If two are required, use the one which needs the most fuel. Otherwise, enough for:

    - Missed approach from MDA/DH at the destination to missed approach altitude, through the whole procedure

    - Climb from missed approach altitude to cruise.

    - Cruise from top of climb to top of descent.

    - From top of descent to where the approach is initiated, through the expected arrival procedure.

    - Approach and landing at the alternate.

    - For helidecks in hostile environments, 10% of the above.

- **Final Reserve Fuel**. This is sometimes called *holding fuel*, and should never have to be used, as it is a final safety margin. It should be:

    - For Day VFR navigating visually, 20 minutes at best range speed, or

    - For IFR or when VFR and navigating by other than visual means or at night, fuel for 30 minutes at holding speed at 1 500 ft (450 m) above the destination in standard conditions, calculated with the estimated mass on arrival above the alternate, or the destination, when no alternate is required.

- **Extra Fuel** as required by the commander or other commercial reasons (it might be more expensive where you are going).

At takeoff, therefore, you should have at least Minimum Reserve Fuel and Trip Fuel on board. *Minimum Reserve Fuel* is Reserve Fuel without the Extra Fuel. *Block Fuel* is Takeoff Fuel plus Taxi Fuel (technically, block fuel is that used for a trip, or part of a trip. It includes fuel for start and taxy, climb, cruise, descent, landing, etc.)

In-flight replanning procedures must include the above, but without Taxy Fuel, and the Trip Fuel would be for the remainder of the flight.

USA

Under 14 CFR (FARs) Part 91.151 - VFR, considering wind and forecast weather, you must *start* with enough fuel on board to fly to the first point of intended landing and, at normal cruising speed, after that for at least 20 minutes. There's nothing to say you have to land with it!

Drums

Drums should not be stored vertically for long periods, because the bungs are not airtight, even though they might stop fuel from leaking out (although it is good idea to stand them vertically for about half an hour before you use the fuel). When the contents contract as the air cools overnight, water inside the rim and collecting around the bung can be sucked in as well, so either store the drums on their sides, with openings at 3 or 9 o'clock, or stick something underneath at 12 o'clock that causes the drum to slant enough to stop rainwater collecting and covering the bungs. Other openings or connections should be protected with blanks or covers, or at least have their openings left facing downwards. Drain plugs, valves, filter bowls, sumps and filter meshes should be checked daily for sediment, slime or corrosion. Always have spare filters.

The reason why long-term storage is not good for fuel (up to two years for drums is the accepted maximum, but some companies reduce this to 1 year) is partly because of daily temperature changes. When it is warm, the fuel expands and some of the vapour-air mixture is driven out. When it gets cold again, the fuel contracts and fresh air is sucked in, to mix with more vapour. As the cycle repeats itself, the fuel inside gradually loses its effectiveness (it loses burn-units). Humidity will mean that water vapour will get in, too, and condense into liquid. Oxygen will also cause a gum to form, which is more apparent when fuel evaporates (the fuel filter is designed to remove it). Keeping the container full will minimise this, which also applies to fuel tanks on machines parked overnight. 100/130 will apparently last for decades, actually longer than the drum, which will deteriorate first, due to the chemicals inside - fuel in drums will interact with the Teflon coating on the inside, if it has fallen off.

Containers should be filled to 95%, and sealed tightly, in a place where the temperature is mostly below 80°F, out of direct sunlight. The 5% airspace allows room for expansion.

Fuel Checking

Each day before flying, and when the fuel is settled, carry out a water check in aircraft and containers (but see below, for drums). Collect samples in a transparent container and check for sediment, free water or cloudiness - if there is only one liquid, ensure it is not all water. The instructions for using water detectors are displayed on the containers. In the Arctic, unless there is a thaw in Summer, separated water will be frozen in the bottom of the drum, and you will only have to worry about that in suspension. Water-finding paste, however, will not detect suspended water, and is as an additional test, not a replacement for a proper

inspection. Oxidised fuel will be darker than normal with a rancid smell. A rotten egg smell indicates fungal activity.

Naturally, only competent and authorised personnel should operate fuelling equipment, who must also be fully briefed by their Company. In practice, of course, refuellers know very well what they're doing, but you should still be in full communication with them. In general, the following precautions should be taken:

- Documentation must reflect the fuel's origins and handling

- Vehicles must be roadworthy and regularly inspected

- Extinguishing equipment must be available and crews familiar with its use

- Always stand out with the refueller, and tell him clearly what fuel your ship needs - you don't know his standard of training, and he could think that all helicopters are jet powered (true!) In other words, never assume the refueller knows what he is doing (if it comes to that, never assume that you know what you are doing!)

- Barrels, when used, should be undamaged and in date (give-aways for this include faded labels). Over long periods, a fungus can grow, which will clog fuel lines. When checking a drum, have it standing for as long as possible, but at least half an hour (although the benefits of this are negated when drums are stored on their side at the fuel cache and you need the fuel in a hurry). Place a block of wood at some point between the bungs, so that dirty fuel is kept more away from the openings and any garbage at the bottom is away from the bottom of the standpipe. Then draw a sample from as far down as you can through a water detector. If you put the standpipe in, block the top with the palm of your hand, and pull it out, you can empty the standpipe into a container to make this easier. Smell the contents - don't trust labels or colours if the seal's broken. Also, get used to the weight - water weighs more and avgas weighs less than turbine fuel. An X on the drum means contamination. Secure it afterwards so it doesn't roll around the landing site

- Run fuel for a few seconds to clear the pipes of condensation and bugs, etc., downstream of the filters

- Keep exit paths clear for removal of equipment in emergency

- The aircraft, fuelling vehicle, hose nozzle, filters or anything else through which fuel passes should be electrically bonded *before* the fuel cap is removed (if you haven't got a bonding cable, at least touch the fuel cap with the nozzle, and keep it in contact with the filler neck when fuelling). With drums, the accepted procedure is drum to ground, drum to pump, pump to aircraft, nozzle to aircraft then open the cap, and the reverse when finished. Be more careful when it's cold, as the air will likely be dry, and airborne snow particles will add their own friction, and static. However, according to NFPA 407, App A A-3-4, if the machine and drum are bonded, they don't need to be grounded. This is because "it does not prevent sparking at the fuel surface" (NFPA 77, *Recommended Practice on Static Electricity*). The National Fire Protection Association is *the* authority on this subject). If you do feel the need to ground anything, salt water is better than permafrost.

  It's not only the movement of fuel through pipes and filters (especially filters) that generates static, but also a fault in some part of the system may apply a voltage to the nozzle. Plastics don't help, and using chamois as a filter is dangerous

- Don't refuel within 100 feet of radar equipment that is operating (even an HF radio transmitting will energise strip lighting). Only essential switches should be operated, with radio silence observed during fuelling

- Avoid fuelling during electrical storms, and don't use bulbs or electronic flash equipment within the fuelling zone. Non-essential engines should not be run, but if any already running are stopped, they should not be restarted until fuel has ceased flowing and there is no risk of igniting vapours

- Brakes or chocks should be applied, but some places require brakes off when near fixed installations

- Take out rescue and survival equipment so if it blows up you have something

The most important thing is daily checks, before flying. If you spill anything, either use a neutralising agent, move the aircraft or wait for it to evaporate.

Note: *Fuel can burn you.* High vapour concentrations irritate the eyes, nose, throat and lungs and may cause anasthaesia, headaches, dizziness and other central nervous system problems. Ingestion (as when siphoning while defuelling) may cause bronchopneumonia or similar nasties, including leukemia and death. If you get it on your clothes, ground yourself before removing any and rinse them in clean water. Spills must be covered with dirt asap.

Otherwise, everybody not involved should keep clear - at least 50m, but see later.

Passengers on Board

You should not normally refuel with passengers on board, especially when the engines are running, but in certain circumstances (i.e. casevac, bad weather, no transport, or on an oil rig) it may be permitted, if:

- Passengers are warned that they must not produce ignition by any means (including switches). They must also stay seated, with harnesses unfastened

- "Fasten Seat Belt" signs are off, and NO SMOKING signs on, with interior lighting to identify emergency exits

- A responsible person is at each door which is open and free from obstacles

- Fuellers are notified if vapour is detected

- Ground activities do not create hazards; the fuel gear should not stop people leaving in a hurry

- ATC and the Fire Authority are informed

- Fire extinguishers are close by

DITCHING

Trivia: The word ditching comes from RAF slang used during World War II - the English Channel was referred to as "The Ditch", in the same way that the Atlantic is referred to as "The Pond", so ending up in The Ditch (voluntarily or otherwise) became known as Ditching.

For our purposes, Ditching is a deliberate act, rather than an uncontrolled impact, although the terms are often used synonymously. A successful one depends on sea conditions, wind, type of aircraft and your skill, but it's the after-effects, like survival and rescue that appear to cause the problems (88% of controlled ditchings happen without too many injuries, but over 50% of survivors die before help arrives).

Of course, the best way out of a ditching is not to get into one, but you can't always avoid flying over water. The next best thing is to prepare as much as possible beforehand, and make sure that the equipment you need is available, and not stuck in the baggage compartment where no-one can reach it. Have you really got enough fuel for the trip? Did you top up the oil or check the weather?

Once under way, flying higher helps in two ways, by giving you that little extra time to reach land, and to allow you to brief and prepare the passengers better. Maintaining a constant listening watch helps somebody know your position, as does filing a flight plan before going.

Sea Movement

It's a good idea to have a basic knowledge, as getting the heading right may well mean the difference between survival and disaster. The maximum wave size for a Bell 47 or a 206 is 3 feet, from trough to crest, which is not particularly large. The 205/212 can cope with 1 foot. Whereas waves arise from local winds, swells (from larger bodies of water), rely on more distant and substantial disturbances. They move primarily up and down, and only give the illusion of movement, as the sea does not actually move much horizontally. This is more dominant than anything caused by the wind, so it doesn't depend on wind direction, although secondary swells may well do. It's extremely dangerous to land into wind without regard to sea conditions; the swell *must* be taken into consideration, although it could assume less importance if the wind is very strong. The vast majority of swells are lower than 12-15 feet, and the swell face is the side facing you, whereas the backside is away from you. This seems to apply regardless of the direction of swell movement.

The Procedure

You will need to transmit your MAYDAY calls and squawks (7700) while still airborne, as well as turning on your ELT, or SARBE. If time permits, warn the passengers to don their lifejackets (without inflating them, or the liferafts) and tighten seat belts, remove headsets, stow loose items (dentures, etc.) and pair off for mutual support, being ready to operate any emergency equipment around (they should have been briefed on this before).

One passenger should be the "dinghy monitor", that is, be responsible for the liferaft. If it's dark, turn on the cabin lights and ensure everyone braces before impact (the brace position helps to reduce the flailing of limbs, etc. as you hit the water, although its primary purpose is to stop people sliding underneath the strap; there are different ones for forward and aft seats).

If only one swell system exists, the problem is relatively simple - even if it's a high, fast one. Unfortunately, most cases involve two or more systems running in different directions, giving the sea a confused appearance. Always land either on the top, or on the backside of a swell in a trough (after the passage of a crest) as near as possible to any shipping, meaning you neither get the water suddenly falling away from you nor get swamped with water, and help is near.

Although you should normally land parallel to the primary swell, if the wind is strong, consider landing across if it helps minimise groundspeed (although in most cases drift caused by crosswind can be ignored, being only a secondary consideration to the forces contacted on touchdown). Thus, with a big swell, you should accept more crosswind to avoid landing directly into it. The simplest way of estimating the wind is to examine the wind streaks on the water which appear as long white streaks up- and downwind. Whichever way the foam appears to be sliding backwards is the wind direction (in other words, it's the opposite of what you think), and the relative speed is determined from the activity of the streaks themselves. Shadows and whitecaps are signs of large seas, and if they're close together, the sea will be short and rough. Avoid these as far as possible - you only need about 500' or so to play with.

The behaviour of the aircraft on making contact with the water will vary according to the state of the sea; the more confused and heavy the swell, the greater the deceleration forces and risks of breaking up (helicopters with a high C of G, such as the Puma, will tip over very easily, and need a sea anchor to keep them stable - as it happens, the chances of any helicopter turning upside down are quite high). Landing is less hazardous in a helicopter because you can minimise forward speed. In fact, if you are intentionally ditching, you should come to a hover above the water first, then throw out the kit and the passengers. Having moved away, settle on the surface. If you can't do that, try for a zero speed landing, which means a steep flare a little higher and sooner than normal - any fore and aft movement on landing may cause rocking. Level off higher, as well. If open, doors will likely be slammed shut anyway.

You need to protect your thumbs throughout the whole process, as undoing a seat belt is a lot more difficult without them. Another tip is to reduce the length of your neck by hunching your head into your shoulders, like a turtle. Be particularly aware that anything happening to the blades will be transmitted through the controls, and may well be painful, or worse, if you get the cyclic in your stomach. At some stage you will be able to do nothing further with the controls, so be prepared to take place your limbs so that they do not flail about.

Keep your knees together, and prepare to use the hand near the exit to get out with, and the other to release the seat belt, but not until the machine is completely under water and the rotors have stopped moving (wait at least 7 seconds). This keeps you in the same relative position to the chosen exit, as well as giving extra leverage if you have to push against anything. Remove your headset, as the cable may keep you in!

Once on the water, hold the machine upright and level using all the cyclic control there is, and use the rotor brake (if you've got one). Then let the aircraft sink. Rolling towards the retreating blade is one consideration, but this will increase the chances of disorientation, although it does ensure that the engine or transmission moves away from the cabin if it breaks free, due to gyroscopic precession. The way out of a submerged cabin is to place a hand on an open window or door, and follow your hand out, so you have a better idea of which way is up. Otherwise, instruct passengers not to leave until everything has quietened down. When you do, take the flotation and survival gear, but keep everyone together (remember that even seat cushions float). Attach the raft to the aircraft until you need to inflate it, as it will sail away downwind quite easily.

Splash, use flares or mirrors to attract attention, but let the rescuers come to you. Don't leave the security of the raft or aircraft unless you're actually being rescued as the downwash or wind will blow them away from you.

If you're in the water, keep moving. You won't be able to see much, because of the swell, so don't attempt to swim unless land is less than a mile or so away. Use your energy to get your back to the swell and keep from ingesting water. Whatever you do, DON'T DRINK SEAWATER - it absorbs liquid and body fluids are used to try and get rid of it, so it gets you twice. Cold makes you give up, so try and keep a positive mental attitude. Except in mid-ocean, SAR will be operational very soon after the distress call, so switch on the SARBE or ELT as soon as convenient, which will also assist a SAR satellite to get a fix on you. Try not to point the aerial directly at rescue aircraft as this may put them in a null zone.

Don't worry if the rescue helicopter disappears for ten minutes after finding you. It will be making an automatic letdown to your exact position after locating your overhead at height. This is where the temptation to use speech is very strong, but should be resisted because this is when the homing signal from the ELT/SARBE is most needed. Speech should only be used as a last resort as, not only will it wear your batteries down, but also take priority over the homing signal. If you feel the need to do something, fire off a few mini-flares instead. Or scream.

Finally, once in the winch strop, don't grasp the hook, because of the possibility of shocks from static electricity. See also the discussion on Immersion Suits, below.

Equipment

This needs to be for aviation use.

RAFTS

Aviation liferafts are designed to vent to atmosphere in case of a problem, rather than into the liferaft itself, as is the case with marine ones (they could inflate in the cabin). As it will float before it's fully inflated, tie it to the airframe (unless it's actually sinking), or a person, before inflating (in fact, it should be tied to at least one person as much as possible). Do this downwind, so it doesn't get damaged against the aircraft. To turn it upright in the water, get downwind, and place the cylinder, which is heavy, towards you. This weight, plus the wind, will help it to flip over. Once inside the raft, protect yourself as much as possible with the canopy, and get the sea anchor out. Buoyancy chambers should be firm, but not rock hard.

LIFEJACKETS

An unconscious person needs 35 lbs of buoyancy to keep afloat, so make sure they are so capable, especially taking a fair bit of wear and tear. Automatically inflated types activate when a soluble tablet gets wet, which is no good in a water-filled cabin, as you will be unlikely to get out of the cabin entrance. Purloining them from airlines is also not a good idea, as they use one-shot jackets. The reason CO_2 is used to inflate them is that it doesn't burn.

IMMERSION SUITS

Immersion suits are useful, but they are not necessarily to keep you warm long-term, that is, to delay hypothermia, although that is part of their function - a good majority of deaths with a suit on occur well within any time needed for hypothermia to even set in. The real danger is inside the first two or three minutes, from cold shock response, which will reduce your capacity to hold your breath, and possibly set off hyperventilation, aside from contracting blood vessels and raising the blood pressure. At temperatures between 5-10°, the average capability for holding the breath reduces to about 10 seconds, if at all.

From 3-15 minutes, the problem appears to be keeping the airways clear - it can be quite frustrating trying to breathe while you're continually being splashed. It's not till 30 minutes have passed in average conditions that hypothermia starts to rear its head, and if you're not wearing a lifejacket, it will reduce your ability to use your arms to swim. Even the method of taking you out of the water can be dangerous if it causes the blood to pool away from the cardiovascular system - whilst in the water, its pressure against your body helps return blood from the lower limbs back to the heart - this support is removed once you are out.

REMOTE AREAS

Because of communication in remote areas, Ops, or someone responsible, must know where you are. If you have to make a forced landing, you must ensure that the Company is notified together with the appropriate ATC, so that overdue action is not set in motion unnecessarily. In the Sparsely Settled Area of Canada you must be able to communicate with a ground station from any point along your route, which possibly means using SSB HF (5680 KHz), unless within 25 nm of your base or an airport. A typical job done in a remote area is *Site Support*.

When leaving passengers in an isolated position, you need to make sure of a couple of things. Firstly, everyone understands the time (and date) of pickup, the location and the method of backup transportation. Also, keep a record of the names, all relevant grid references, etc. It's a good idea to leave a map with all possible routes and stopping points listed on it so the guys back at base will at least some idea of where to start looking. Keep in mind the recovery problems should the engines fail to start after a shutdown; always position as close as possible to a track or road to save trouble later (engineers like being near a pub as well, if you can manage it). The track or road will also help as a line feature to make your way back with.

Don't let your fuel get too low - it's usually delivered to accurate GPS coordinates, which may be on top of a frozen lake so the drums will sink in Spring and not be there when you want them, or be covered in snow, and you wil lhave to dig for 20 minutes, assuming you've found the right place! Either that or Ops may have written them down wrongly. My point is that the added stress of looking for fuel that isn't there when you're short anyway is not what you need. Check where it is before you start work! Certainly, verify your fuel cache sites before passing your PNR!

Assuming your passengers don't carry too much baggage, you should be able to carry a few home comforts, such as a tent, a stove that runs on aircraft fuel, high-calorie food and a sleeping bag rated for the relevant temperatures. Keep it out of the aircraft when refuelling, so you don't get left with nothing if it catches fire.

If you're forced down, the same principles of passenger preparation for landing apply as for ditching (see above). Having arrived on the ground, the first task (if necessary) is to assist survivors and apply First Aid, after turning on the ELT or SARBE if you have one, and the second to provide shelter (once the ELT is on, leave it on, as that will make best use of the batteries). The absence of food and water should not become a problem for some time if everyone's had their breakfast - even in the Arctic, in Summer, there's plenty of water around, but you would still be wise to boil it first, for at least 5 minutes, as cold does not kill germs. Try not to eat or drink at all for the first few hours, and divide whatever you have into equal parts. When you do eat, go slowly and eat small amounts of food. It's generally best to avoid mushrooms, as well.

Consider using the aircraft for shelter if it hasn't burned away, and has actually stopped bouncing. In the Arctic, move the wreckage if you can to the highest point around, so you can be seen more easily. Maybe take the cowls off and use them as reflectors. Don't wander too far away from it, and ensure that everyone stays within sight of each other. Use remaining fuel for light and heat as necessary (fuel must be warm before it will light) and maintain a positive mental attitude.

The best cure for hypothermia, when your body loses more heat than it produces, and your organs lose their ability to function, is to use blankets and lukewarm sweet drinks. Direct heating, as with hot water bottles, will only serve to open up surface blood vessels and take heat away from the core organs, where it's most needed. Victims may also vomit, so give them nil by mouth, even when they are alert. They may have altered levels of consciousness, so handle them carefully. Hypothermia happens quite slowly, and arises from cold and wind, poorly insulated or wet clothing, prolonged immersion in even warm water, and fatigue (in water, heat is conducted away 25% faster than in cold air). Shivering and grogginess are among the early symptoms, allied with poor judgement and muddled thinking. As it gets worse, the shivering may stop and the attention span will reduce, together with shallow breathing and a slow, weak pulse. Unconsciousness and little or no breathing signifies the full thing, with dilated pupils.

Although direct heating is not recommended for hypothermia, it is a better solution for frostbite than friction (for example, cupping a frozen ear with your hand will have a better effect than rubbing it). The only reason you would hold snow against a frostbitten part of your body is to relieve any pain caused by warming - it will not help with the original condition! Also, the fact that a liquid is not frozen doesn't mean it will help, either.

Ground-Air Visual Signals

Survivors can communicate with SAR aircraft visually by making signals on the ground. They should be at least 8 feet high (or as large as possible) with as large a contrast as possible between the materials used and the background.

Need Assistance	V
Need Medical Help	X
No	N
Yes	Y
Going This Way	←

Rescue units can use these (mostly double symbols):

Operation Complete	LLL
Found all personnel	LL
Found some personnel	+ +
Cannot continue - going home	XX
Split into different groups in directions indicated	← →
Aircraft in this direction	→ →
Nothing found but continuing	NN

Air-Ground Visual Signals

Show your understanding of the signals above by rocking your wings in daylight or flashing your landing lights twice at night (or nav lights if you haven't any).

Droppable Containers & Packages

Those containing survival equipment should be coloured (with streamers):

Red	Medical Supplies/First Aid
Blue	Food and Water
Yellow	Blankets/Protective Clothing
Black	Miscellaneous (stoves, shovels, etc.)

Use a combination if the goods are of a mixed nature.

EMERGENCY EQUIPMENT

Every aircraft carries a First Aid kit that conforms to whatever standards are relevant, and is certified by an engineer. Lifejackets are commonly under the relevant seats when carried, or on the rear parcel shelf, and liferafts should be securely stowed but easily accessible.

ELTs are supposed to come on automatically, and they generally do if they are attached to fixed wing aircraft, assuming the batteries are kept up to scratch and they are checked regularly, but, with helicopters, there are fewer guarantees that this will happen. For a start, there's a lot more vibration, and there are less places to attach it, as they should be fitted as far aft as possible, aligned fore-and-aft so the shock forces activate the G switch properly. Where it's fitted in the cabin, it's often switched off so it doesn't get kicked or bashed and set off accidentally, which is why it's a good idea to include switching the thing on as part of your emergency checklist on the way down. Not all military helicopters monitor 121.5, as it's primarily a civilian emergency channel, so get one with 243 Mhz as well, which is where they mainly hang out.

Make sure the survival kit can be easily opened one-handed with cold fingers! Talking of which, this item should also be inspected regularly, as you don't want any nasty surprises when you come to use it, and find that someone's pinched the chocolate, or the matches. As space is limited, food should be of a lesser priority than firemaking and signalling devices, and drinking water, or purifying tablets, at least, and anything specially required for the area you are in, but you still need it (power bars and chocolate carry a good bang for buck). Try to carry it in your pockets, or in a place you can get it in a crash, because Murphy's Law will dictate that the baggage compartment is underneath the hull.

Your local regulations should give you a good list of what is required in a survival kit, but here are a couple of extra items to consider: Magnesium is great for starting fires, as is masking tape (or duct tape in N America), which can also be used for strapping up wounds, etc. Heavy duty garbage bags are great for keeping the rain off, and you may need a licence to carry a weapon, if one is included.

Try not to open your rations or water on the first day, but split them up into equal portions according to how long you think you will be stranded. When you do use them, do so in small amounts, chewing slowly.

ICING

Ice adversely affects performance, not only by adding weight, but also altering the shape of lift producing surfaces, which changes your stalling speed - autorotation could therefore be a lot more interesting than normal (the US Army found that half an inch on the leading edge reduces your lifting capacity by up to 50%, and increases drag by the same amount) - if your engine stops, you could really fall out of the sky!

Zero degrees is actually the point at which water becomes supercooled and capable of freezing. Airframe icing happens when supercooled water droplets strike an airframe below that. Some of the droplet freezes on impact, releasing latent heat and warming the remainder which then flows back, turning into clear ice, which can gather without noticeable vibration. On the ground this can mean ground resonance, and bits of ice flying off rotor blades. In flight, the extra weight and drag could cause descent and improper operation of flying controls. So - it's a good idea to avoid icing conditions but, in any case, you shouldn't go if you haven't got the equipment, which naturally must be serviceable. The "clean aircraft concept" means that nothing should be on the outside that should not be there, except, perhaps, for deicing fluid.

All ice should be removed from critical areas before takeoff, including hoar frost on the fuselage, because even a bad paint job will increase drag, which is relevant if you're heavy, and hoar frost will have a similar effect. Deicing details should be entered in the relevant part of the Tech Log, including start/end times, etc. The critical areas include control surfaces, rotors, stabilisers, etc.

The ability of an object to accumulate ice is known as its catch efficiency; a sharp-edged object is better at it than a blunt-edged one, due to its lesser deflection of air. Speed is also a factor. Due to the speed and geometry of a helicopter's main rotor blades, their catch efficiency is greater than that of the fuselage, so ice on the outside of the cabin doesn't relate to what you might have on the blades. In fact, Canadian Armed Forces tests show that you can pick up a lethal load of ice on a Kiowa (206) rotor blade inside 1-6 minutes, although it's true to say that 206 blades, being fairly crude, don't catch as much as more sophisticated ones, such as those on the 407. It's also true that some helicopters, such as the Sikorsky S61, will not take ice on the main or tail rotor blades down to about -1°C, due to friction.

The rate of accretion is important, not the characteristics of the icing, although clear ice is definitely worse than rime ice, since the latter contains air bubbles and is much lighter and slower to build - it also builds forward from the leading edge as opposed to spreading backwards. Variations on clear ice are freezing rain and freezing drizzle, both of which have larger droplets and are caused by rain, snow or ice crystals falling through a layer of warmer air at lower altitudes. However, the latter's droplets have a much higher water content.

Although aircraft are different, expect icing to occur (in the engine intake, anyway) whenever the OAT is below 4°C. Otherwise, it can form in clear air when humidity is high - anti-icing should be switched on in advance. Pitot head, static and fuel vent heaters should be on whenever you encounter icing, plus anything else appropriate.

You need warmer air to get rid of ice effectively - just flying in clear air can take hours. Climbing out is often not possible, due to performance or ATC considerations, and descending has problems, too - if you're getting clear ice, it's a fair bet that the air is warmer above you, since it may be freezing rain, which means an inversion, probably within 1000 feet or so, as you might get before a warm front. In this position, landing on your first attempt becomes more important as you are unlikely to survive a go-around without picking up more of the stuff.

Ground De-icing

Use either soft brushes, fluids, or a combination. Priorities are control surface hinges, engine intakes or static ports. Some manufacturers, however, (e.g. Bell) don't recommend using fluids at all because of the possible effect on the bonding of composite materials. They are also very efficient degreasers. Warm water can be used, but I'm not so sure about that - I know if you want to freeze water quickly, you put it in the freezer hot.

General Precautions

Deposits must be swept away from hinge areas and system intakes, and the sprays themselves should not be directed to them, since the fluid may be further diluted by the melting ice it is designed to remove, and may refreeze. It may also cause smearing on cockpit windows and loss of vision during takeoff. Afterwards, confirm that flying and control surfaces are clear and move over their full range, and intake and drain holes are free of obstructions. Jet engine compressors should be rotated by hand to ensure they are not frozen in position.

RECORDING OF FLIGHT TIMES

Flight times in personal logbooks are from first movement under power until rotor rundown. Those in Tech or Journey Logs, by contrast, are from skids up to skids down, sometimes entered in decimal hours. It's common practice, when several flights are made per hour without closing down, to record the first takeoff and last landing and note the actual airborne time in between.

There are many ways of doing this, the most accurate using a stopwatch, but there is an unofficial and widely used practice (by arrangement with your Inspector), when doing lots of sectors between engine starts, of using two thirds of the total time between first takeoff and last landing. Accountants love it, but engineers don't, as they regard the wear and tear as still taking place. Too much of this will really play havoc with servicing schedules (and profit and loss figures) as parts will wear out quicker than anticipated, despite the 'fudge factor' allowed when setting up maintenance requirements.

PASSENGER SAFETY

How to handle passengers in general is very much a matter of Company policy. Some like to be spoken to, some don't, but there are some small attentions you can give without being obtrusive. Just going round checking seat belts and doors helps (never trust a passenger to shut doors properly), as is a look over your shoulder before takeoff and occasionally during the flight.

People new to flying are fairly obvious, and they may not appreciate such commonplace occurrences (to you, anyway) as noise, turbulence, pressure changes or lack of toilets. However, you are responsible for the safety and well-being of your passengers. You are supposed to brief them before every flight, or at least take all reasonable steps to do so, although what you can do with the nose of your helicopter in the side of a mountain and your hands on the controls is a bit different from what you can do on the ground with a bit more time, so try and get as much done as possible beforehand.

A lot depends on what your passengers are doing when they arrive - if you're going to shut down, tell them to stay seated until everything stops (it helps to explain why you have to sit there for 2 minutes). If it involves a running disembarkation (other than Pleasure Flying), one passenger should operate the baggage door and do the unloading. Everyone else must leave the rotor disc area.

Nobody should enter the area covered by the main rotor disc without your permission (indicated by "thumbs up" during the day, or a flash of the landing light by night). Movement in and out of it should be to the front or at 45° to the longitudinal axis, ensuring that all movement is within your field of vision. Additionally, no movement should be allowed during startup or rundown (due to blade sailing) and nobody should approach the rear of a helicopter AT ANY TIME (unless it's a Chinook). You can help by landing so that passengers have no choice but to go forward, but watch the doors aren't forced against their stops if the wind is behind you.

Tip: Do not reduce the throttle to ground idle when passengers are getting in and out, so when one of them decides to run round the back (they will), you can lift into the hover to move the tail rotor out of the way.

Transistor radios, tape recorders and the like should not be operated in flight as they may interfere with navigation equipment. If you don't believe me, tune to an AM station, as used by ADF, on a cheap radio and switch on an even cheaper calculator nearby - you will find the radio is blanked out by white noise. In fact, the radiations from TVs and radios (yes, they do transmit - how do you think the TV detector people find you?) come within the VOR and ILS regions as well. Cellular phones are dodgy, too, but when you're flying, you are further away from a cell, so the phone pumps up the power to keep in touch. Too many phones doing this really will interfere with the instruments.

Anyway, you, as commander, are responsible for ensuring that all passengers are briefed, or have relevant equipment demonstrated. Where you work with regular passengers, say in a corporate environment, you can probably do away with a briefing for every single flight, and just use a briefing card as a reminder.

Tip: On the other hand, in a remote bush camp, for example, you could get everyone together (including the cook) and do them all in one go, once a week. Naturally, some will complain that they don't need to do it then, but you could explain to them that the only way out at the end of their tour is by helicopter, and a briefing then will take more time, which is just what they need when they have a flight to catch.

Use this checklist as a reminder that you've covered everything:

- Your authority as Commander

- Methods of approach, in particular avoiding exhausts and tail rotors - if nearby aircraft have engines running, it could mask the sound of a closer one Pitot tubes are especially sensitive (and hot!). Children should be kept under control. Wait for signal from pilot. Used crouched position in pilot's view. Take off loose objects, hats, etc.

- Loading of baggage and items that must not be carried. Bear scares (pepper sprays) must not be in the cabin. No objects carried above shoulder height - and horizontally in any case. Long equipment should be dragged by one end. Don't throw cargo

- Opening and closing cabin doors and their use as emergency exits. Not leaving seat belts outside. Where not to step and what to hold on to. Sharp objects must be handled carefully when working with floats

- Hazards of rotor blade sailing and walking uphill in the disc with rotors running

- When they can smoke (not when oxygen is in use!)

- Avoidance of flying when ill or drunk - this is dangerous to themselves, but if they are incapable next to an emergency exit, others could suffer too

- How to use the seat belts and when they must be fastened

- What not to touch in flight

- Loose articles, stowage (tables, etc.) and dangers of throwing anything out of the windows or towards any rotor blades

- Use and location of safety equipment, including a practical demonstration (if you intend to reach a point over 30 mins away from the nearest land at overwater speed, do this with the lifejacket)

- The reading of the passenger briefing card, which should be of at least Letter or A4 size, so it doesn't get lost in a pocket. It should also be as brightly coloured as possible, so it catches the eye. Particular things to place on this card that always seem to be forgotten include instructions not to inflate lifejackets in the cabin and full door opening instructions (don't forget any little bolts that may be about)

- The brace position (including rear-facing seats). If you ever have to give the order to adopt it, don't

do it too early, otherwise the passengers will get fed up waiting for something to happen and sit up just at the point of impact

- Landing areas should be clear

- No-one in the cabin when slinging, or on the longline

- How long the flight will be, and how high you will be flying, the weather, etc.

NIGHT FLYING

Night flying can be pleasant - there's less traffic, you tend not to go in bad weather and the air is denser, so the engine and flying controls are more responsive (if they're heavy, the instructor has his hands on as well). However, the tendency is not to allow single-engined night flying on Commercial Air Transport, but occasionally positioning may take place with the pilot only on board. This doesn't make it any safer, but at least reduces the number of awkward questions. Don't forget there is no VFR at night, so you have to go under IFR (not the same as going IMC).

Searching for an overdue aircraft in low light conditions causes lots of problems, and route planning should take account of this. Otherwise, it's much the same as for day, except the air is a different colour, though there are some aspects that demand some thought. Plot your route on the chart in the normal way, but navigate with electronic aids or features that are prominent at night, such as town lighting, lighted masts or chimneys, large stretches of water (big black holes), aerodromes, motorways, etc. It's often convenient to go from lighted area to lighted area, but this depends on what you're flying over. However, the easiest way to get around is to know the area you are flying in. Get used to it by day as much as possible, and establish some good safety altitudes.

Helicopter landing sites should be checked out in daylight on the same day as they are to be used at night. Preflight checks should allow for night flying - carry a torch, and 2 landing lights are preferred, although my own preference is not to use them if there is adequate other lighting around, as they tend to shine up into the cockpit. Permission to enter the rotor disc is given by flashing the landing lights.

Hovertaxi higher and slower than by day, making no sideways or backwards movements.

Schermuly Flares

Great care should be exercised in pointing the Schermuly flares to a safe place at all times (which is admittedly a bit difficult when they're fitted and the fuelling truck pulls up right alongside them). The flares should not be armed at this stage, but at the holding point immediately before take-off and disarmed at the same place after final approach. They should also be disarmed after reaching cruising altitude. The maximum useful height for discharging a flare is around 1800 feet. Its burn time is 80 seconds, during which time it will fall about 1500 feet. Therefore, having established autorotation after an engine failure at night, the first flare should be discharged immediately, or on passing through 1800 feet, whichever is later. Don't bother doing it before this, as they will be useless. Due to the way the switches work, and depending on the height at which your engine stops, you may not be able to set off more than one flare before landing, but, if possible, the second should be discharged between 800-1000 feet agl.

In autorotations at night, use a constant attitude, at whatever speed is comfortable, to keep the beam from the landing light in the same position on the ground, because otherwise it will shine up into the air when you flare.

WINTER OPERATIONS

Although colder air means there's less danger of exceeding temperature limits, there are hazards, including freezing precipitation, low ceilings and cold temperatures. Rapid changes in these are typical, and you can get weathered in for days at a time, so don't forget your rations!

There are reasons for minimum operating temperatures:

- Servo seals shrink when cold soaked to about -32°C, and the hydraulic system could leak badly.

- Elastomerics do not like to be stretched too far when cold soaked, which is why you shouldn't move the controls for a short while when starting

- Cockpit displays can stay blank until they warm up, so it's difficult to see how the start is going!

The Weather

In the Frozen North, the best conditions are in late winter or early spring, with one of the major problems being darkness. Once the snow is down, the air is dry and it can stay clear and cold for long periods, so you can usually ignore fog and the rest until it gets a bit warmer.

Above the 60th parallel, don't expect the weather to behave rationally at all. For example, further South, the East wind is responsible for bad flying conditions, but up there the West wind is the culprit, as well as large swings between low and high pressure which will often bear no relation to what the weather is doing (so don't rely on cloud shadows over the ground as an indicator of surface wind speed). Aside from barometric changes, look out for wind shifts, which will bring changes in wind speed and amounts of blowing snow and less visibility - even a difference of 100 feet in elevation can mean the difference between snow or not. Temperature changes often mean bad weather is approaching from the North - if it drops, expect ice crystal fog, which is the low level equivalent of contrails made at high altitude, and created by air disturbance, which could actually be from the aircraft itself. Rising temperatures will produce melting and poor visibility. The chill factor from rotors can reduce the ambient temperature by several degrees.

When it gets to below -20°C or so, contact gloves will stop your skin freezing when it comes in contact with cold metal, which is a more efficient conductor of heat than air is. You may also need sunglasses. Always dress properly - in a forced landing it could be that the clothes you wear will be the only protection you have. *This includes customers!* Most pilots don't go too far away from base when it's below -35°C because it's dangerous if you end up camping out for the night (that's assuming the machine's operating envelope allows it in temperatures that low anyway).

Being cold when you are actually flying is a Flight Safety hazard - metal foot pedals will conduct heat away from your boots very quickly. Extra time for planning should always be allowed and the pre-flight inspection should include you - being improperly dressed and making a series of short exposures will fatigue you more quickly, especially when the clothes you are wearing are bulky and awkward to move in. Maintain blood sugar levels as more calories are consumed in the cold (you need at least 3000 calories when it's cold). If the air is very dry (like in the Arctic), you will lose fluids more quickly through the usual ways, but especially breathing. Losing 10% causes delirium, and a 20% loss is fatal. You could try and eat snow, but the conversion to water takes more energy, so melt it first. DO NOT skip breakfast or that warm meal before you go!

Preserve your machine's heat as much as possible on the ground, by covering vital areas as soon as possible after landing, and not opening and closing doors too much, etc. It's very important that it does not get so cold that it won't start again (*core lock* occurs when the outside of the engine shrinks enough to stop the wheels turning inside), so you might consider starting up every couple of hours or so, which will both use fuel and battery capacity - certainly, in the average car, it takes about half an hour's driving to replace the energy taken by one start, and it's the same with a helicopter - a depleted battery will sooner or later result in an expensive hot start. At the very least, use a battery warmer, or even remove the battery and keep it warm - at -30°C, your battery will have less than half its available power to start an engine that needs 350% more effort to get going. If you see fan heaters around the helipad, they are for putting under the covers to keep the engine and gearbox warm (all night), and inside the cockpit to keep the screens warm and stop them frosting over (covers are all very well, but they scratch - and rotors covers just freeze to the rotors, so most people don't use them for anything less than an overnight). Once, the aircraft is out of the hangar, keep it out.

Special attention should also be paid to the following:

- That correct oil and grease is used and special equipment (like winter cooling restrictors) is fitted to keep engines warm. For Bell 206s, at least, below -40C, your oil must meet MIL L7808 specifications, and you will need fuel additives in all fuels other than JP4 below -18C. **Note**: It has been found that when visible water is present in jet fuel containing anti-icing additive, the additive will separate from the fuel and be attracted to the water. After a certain amount, thought to be about 15%, the density of the new liquid changes so much that it is not identified as water, and will therefore pass through water filters, and will also not be detected by water finding paste, which is not, in any case, meant to detect water in suspension. Where the ratio becomes 50%, as much as 10% of whatever is going through the filter could actually be water, which is very likely to get to the engine, since the filters on the airframe itself are not as restrictive.

- Use deicing fluid if possible - scrapers do not leave pretty results. Fluid, if it's thick enough, helps prevent further ice forming, but don't forget to fit engine blanks, etc. before using it. Bear in mind that deicing fluids are also efficient degreasers, particularly alcohol-based ones. No deicing fluids have been approved or tested for Bell products.

- You have proper tie-downs and pitot/engine covers, static vent plugs, etc.

- That windscreens are defrosted (if you use a mechanical heater, keep moving it around, or it might melt the perspex). Don't forget to have a cloth handy for wiping the windscreen from the inside when it mists up.

- That heating systems are working properly and don't allow exhaust into the cabin (if you get regular headaches, check for carbon monoxide).

- De-icing and anti-icing equipment is working and that all breather pipes, etc. are clear of anything that could freeze.

- That the aircraft has not been cold soaked below minimum temperatures. If so, there are particular (and tedious) ways of starting the machine again, which essentially involve preserving the heat from repeated attempted starts so the engine compartment can warm up, with a ten-minute gap between each, removing and replacing engine blankets every time. Just in case you were wondering, *cold soaking* occurs when the aircraft, and fuel, becomes much colder than the ambient temperature, which can happen over a cold night or at high altitudes, and it is a problem because heat is conducted more quickly away from precipitation, making ice formation easier.

- That all frost, ice and snow has been removed, particularly on lift-producing surfaces. If you leave hoar frost on the fuselage (only if it can be seen through), beware of flying into cloud where more will stick. It *must* be removed from where its dislodgement could cause ingestion.

- Check particle separators as water seepage may have frozen inside the engine, resulting in abnormally high N_1 and TOT readings.

- That the skids are not frozen to the ground, so you don't get dynamic rollover. On a solid surface, you might be able to rock it using the tail. Otherwise, use pedals with a little collective just before takeoff, but be careful in a big machine.

- Unstick windscreen wipers and moving parts (including rotors and propellers) by hand, or you will strain the motors.

- That control linkages and movement are checked.

- That pitot heat is checked by hand - don't just accept a flicker on the ammeter.

- Water drains are not frozen.

- That carb heat operation is checked.

- You always have a map handy (they don't need batteries)

- Use short flight plans. If you flight plan ten different legs over a day's flying, SAR will not start looking until the flight plan is up.

- Carry a small tent so you can hide from the wind when you are sitting waiting.

- Carry cash - not many ATMs up there!

Another contribution from Harald on the above subject:

I have done extreme cold operations for the last 10 years, and it is something someone starting out in Canada certainly will see some of, particularly flying freeze up and break up in small communities such as Moosonee in Ontario, or Theresa Point in Northern Manitoba, as well as my now day to day job of flying the North Warning System in the Canadian arctic (along the 69th Parallel). Also some hydro work in Saskatchewan and Manitoba tends to be around the -40 mark (on both scales!) and requires some special care.

- **Servicing**. Take note that your helicopter needs special oils, usually MIL-L-7808 grade for very cold temperature, to be viscous enough.

- **Covers**. Put ALL covers on the machine, including body and blades. Being lazy and thinking that it won't affect your blades may give you a nasty surprise when a mild night with a very cold morning follows. Ask how I know that warm water is the best way to clear ice of rotor blades, and that it is difficult to reach all the way to the tip when you don't have access to a ladder...

  With regards to covers, do NOT let the customer help remove them, as it gets very expensive very fast, and windshields get extremely brittle in temperatures below -40.

- **Heaters**. Ensure that you have little buddy heaters. These are the little car heaters you can put in the cabin (facing the collective, the throttles get very stiff), near the transmission and near the engine, to keep some oils near 0 degrees. If your company operates a lot in extreme cold areas, it is very likely that it has a Tanis kit installed already.

  If you are operating away from base during cold weather (-30 and colder), with no means of plugging in heaters, and you are not flying continuously, make sure that the helicopter gets started every 30 to 45 minutes and ran for a while to get warm, and to charge the battery. There is NOTHING fun in being stuck in the bush with no battery to crank the engine.

- In extreme weather when the machine is parked outside, check on the machine to make sure it is not accumulating lots of ice! In one situation where my colleague (and his engineer) was stuck at a radar site for a few days, they ended up having to literally BEAT the ice of the machine every 6 hours for 2 days, to prevent damage. In freezing precipitation the weight builds VERY fast...

- Remember that pilots need comfort too, and to dress for the situation at hand, but sometimes you have to compromise. I find that the Canada Goose parka is the best thing ever, but it is so bulky that it is impossible to wear and work in, but if you have to hang out for ANY length of time, it is a perfect companion. Dress for the weather outside the cabin, as you can open windows and turn heaters down, but if you don't have enough clothing on and the heater breaks you will be extremely cold VERY fast, and it is tough to have to scrape the inside of the helicopter windshield to see out.

- **Remoteness**. It is impossible to not mention something about the distances. You did mention it earlier, about the vastness of Canada, but until you see the Arctic, you truly can't comprehend it. You can go for 200 miles without seeing anything but a caribou and a Muskox, or a polar bear. The distances are huge, and there is NO, ZERO, NONE fuel available, unless you have caches. So everything has to be planned to worst case scenario, and if there was ever a situation that required maximum fuel, the Canadian Arctic in the wintertime certainly is that.

- **Safety**. There is wildlife around, and anyone becoming a pilot in Canada should get a Possession **and** Acquisition License (PAL), as you may have to carry a shotgun to scare a bear away, or even have to kill one if the situation warrants. There are countless situations where bears have trashed a helicopter, smashed windshields, eaten seat cushions and survival rations etc., or broken antennas just because they are curious. You have to be able to either scare the animal away or protect yourself, and the only real viable option against polar bears are shot guns with slugs. Bear spray, and bear bangers simply don't work.

- Observe aircraft limitations. Some may have a start limit, other may have an operating limit, such as the Bell 212 - you cannot fly below -40 (on both scales), as there are no performance charts below -40. Transport Canada has recognized that is a limit.

Static becomes problem when it's cold, as snow and air can be very dry and therefore good electrical insulators - a helicopter can retain its normal static charge quite efficiently when landing on snow (before refuelling, remove your survival kit, so if it blows up you've got at least something to wear and eat after you've warmed your hands in the fire).

When possible, the first start of the day should be an external one. With a turbine in cold weather you can expect a lower achieved N_1 before light up with abnormally high TOT peaks, eventually settling down lower than normal. Oil pressure will be slow to rise, but high after starting - do not go above ground idle until pressures are in the green and will stay there as you increase the throttle. Temperature, on the other hand, will be very slow to rise at all, and you want the transmission to be at least indicating something, which will mean the engine oil is OK as well, as it gets hot quicker. Allow the electrics to warm up, too - even the knobs can get brittle.

Don't wind up to flight idle too quickly in case you spin or yaw on the pad (the cyclic should be central), especially if there's an engineer on a ladder doing a leak check (be careful with rotor brakes as well). If the machine has been frozen to the ground, one skid may come free first and cause dynamic rollover. If it has not already been freed, pull collective until ready to lift and crack it free with a little controlled pedal movement, though on a big machine you might want to use engine torque for the same effect, otherwise you might bend the tail boom. You could also try gently circulating the cyclic, or pulling down on the tail boom before starting if the machine is light enough. Taxi slowly with caution if the taxiways are clear of snow. If not, taxi higher and slightly faster then normal to keep out of the resulting snow cloud. If you have wheels, act as if you have no brakes.

Tip: If you can manage to start up downwind, the heat from the exhausts will be blown over the cabin and may be useful for clearing frost off the windscreens.

Marshallers should be well clear and move slowly themselves. If the heater is required to be off in the hover, ensure the blower is on, to help clear the windscreen.

Snow

An appreciation of snow can be useful, particularly when mountain flying. It can be hazardous in many ways:

- Reduced visibility when snow is falling. Wet snow sticks to the windshield and creates icing hazards. Colder, falling snow, is better than rain because it doesn't stick or distort your vision, and there is less fogging on the inside

- Snowcloud when landing and taking off (see below)

- Illusions created in featureless terrain and flat light (see *Whiteout* below)

- A solid stable landing site can change quickly to a weak unstable platform, especially when wet. In Spring, it can change from being as hard as rock into mush in about 20 minutes. As skids can sink in at either end, don't let the passengers out unless the belly is in the snow, and the blades are stopped, or are running full speed, with a little pitch pulled - that is, hold it light on the skids with a little power

- Be very aware of the snowball on approach and departure in fresh snow. Always land next to something dark *close to the machine*, like a rock or bush, or a fuel drum or two. Maintain contact with it until you are either able to rise above the snowball, or transition forward from it

See also *Heliskiing* in Chapter 5.

Whiteout

This is defined by the American Meteorological Society as:

> "*An atmospheric optical phenomenon of polar regions in which the observer appears to be engulfed in a uniformly white glow*"

That is, you can only see dark nearby objects - no shadows, horizon or clouds, and you lose depth perception. In other words, you are unable to distinguish between the ground and the sky - the snow-covered surface cannot be detected by the naked eye because of the lack of normal colour contrast.

Whiteout typically occurs over unbroken snow cover beneath a uniformly overcast sky, when the light from both is about the same. Blowing snow doesn't help, and it's particularly a problem if the ground is rising. In fact, there are several versions of whiteout:

- *Overcast Whiteout*, which comes from complete cloud cover with light being reflected between a snow surface and the cloud base. Perspective is limited to within a few feet, but the horizontal visibility of dark objects is not materially reduced.

- *Water Fog.* Thin clouds of supercooled water droplets contacting a cold snow surface. Horizontal and vertical visibility is affected by the size and distribution of the water droplets.

- *Blowing Snow.* Winds over 20 kts picking up fine snow from the surface, diffusing sunlight and reducing visibility.

- *Precipitation.* Small wind-driven snow crystals coming from low clouds with the Sun above them. Light is refracted and objects obscured caused by multiple reflection of light between the snow covered surface and the cloud base.

Once you suspect whiteout, *immediately* climb or level off towards an area where you can see things properly. *Do not be tempted to go on instruments.* Better yet, put the machine on the ground before you get anywhere near whiteout conditions. When operating below a treeline, don't attempt to transit an area you cannot see across such as a frozen lake, or even open fields when the visibility is way down. Fly the edge of lakes or fields along the tree line. If you need to turn back the way you came, do a high hover turn next to a line feature which you can keep in sight at all times.

Flat light is a similar phenomenon to whiteout, but comes from different causes, where light is diffused through water droplets suspended in the air, particularly when clouds are low. *Brownout* comes from blowing sand or dust.

Taking off

In snow, assuming you are not stuck to the ground, the accepted takeoff method is the towering type (i.e altitude over airspeed), because a normal one may produce a large snow cloud to blind everyone, plus a failed engine. However, you need to blow as much loose snow away as possible with a little application of collective before the takeoff proper - pull enough power to get the snow blowing while keeping enough weight on the ground so the machine does not move. Leave the power on as long as you need to get good visual references, which could take up to a minute.

Note: Look out of the front - ignore the mirror! This helps you catch any rollover tendency. Also, take off *slowly* and *carefully* - your skids may have become frozen to the ground, or snagged under snow cover, or a heavy passenger getting out may have changed the C of G!

Once good reference is established, use altitude over airspeed to stay out of the recirculating snow during the rest of the deparyure procedure.

If a white-out happens, apply maximum collective for an immediate climb and forward cyclic (i.e. no hover), keeping the ball centred and using the A/H if necessary, but the real key is keeping a visual reference, preferably inside the rotor disk. If you have wheels, exercise the gear once or twice to dislodge slush, etc. that may have stuck to the legs, to stop it freezing.

When the snow cover is light, say under 5 mm, and dry, you could use a rolling technique that keeps you ahead of the snow cloud. Again, before you start, apply power to blow as much away as possible so you can obtain some reference. When you are ahead of the snow, lift into the air, accelerate to normal climb speed and follow the normal profile.

Note: Use this when the snow cover is light (less than around 5 cm), and dry. Otherwise, deep snow could strain the landing gear, especially wheels.

With a piston engine, use carb heat regularly and check frequently for carb icing. Have carb heat fully on or off, but not on for prolonged periods - it increases fuel consumption markedly (see *Engine Handling*).

The Cruise

Mountain wave clouds can be loaded with heavy ice at remarkably low temperatures (remember that low pressures and low temperatures will cause your altimeter to read high). When using anti-icing, take into account the inaccuracy of the temperature gauges, so if you must turn it on at 4°, and the temperature gauge is only accurate to within 2, start thinking about it at 6°. Wet and sticky snow has more chance of icing, and is associated with low visibility, which would indicate that you shouldn't be flying anyway. Luckily, light powdery snow tends not to accumulate, but will still give you the leans. Whatever you fly in, make sure the baffles are fitted. Visibility, by the way, includes the inside! When it's very cold, water vapour (from clothes, breath, etc.) will freeze on the windscreen, so warming up the machine before passengers get in will help a lot.

Navigation

Sun Tables are used for resetting your DI in the Arctic, with true sun bearings taken every 20 minutes or so (assuming you can see it), based on the fact that we know where the Sun will be with reference to True North for a given time, date, latitude and longitude. Having obtained the local time, look in the tables for the Sun's bearing, point the nose towards it and set the DI to True North. The two types of navigation used are *True North* and *Grid North* (to find Grid North, add your longitude to True North, and vice versa). In True North navigation, headings have to be measured from your point of departure, using its longitude as a base line. Every time you cross a longitude, add a degree going East, and subtract going West, so if you cross 10 longitudes enroute on a heading of 090°T, your return heading will be 280°T, not 270°.

Tip: Another way to determine true heading is to ask your GPS (if it is capable) for the true bearing to an NDB. Then subtract your relative bearing for the true heading.

Many pilots drop dye balloons en route so they can find their way back. Others fly low enough to create a disturbance in the snow with their downwash.

Landing

Landing Sites should be selected with a view to pulling out of a resulting snow cloud if necessary. That is, you need escapes, and *enough power beforehand*. That is, you neeed enough OGE performance to keep you out of trouble. Also, ensure any anti-icing systems are working and switched on.

As with landing on mountains, there are various schools of thought about landing on snow. One is the zero/zero method (zero speed, zero height), for which you carry out a normal approach, using a constant attitude with minimum changes, losing translational lift at the last minute, which requires some timing. This is for when you

do not have enough HOGE performance, and have to fly fast enough to keep ahead of the re-circulating snow and complete a no-hover landing before it surrounds you.

Aim to keep going forward and downward until a few inches above the snow, so the downwash (and snow cloud) is always behind you, using the aircraft shadow, a smoke grenade or the landing light to provide texture to the surface. Even a fuel drum makes a good visual reference (one good trick is to use a dark-painted stake with a flag on it - the flag makes the stake behave like a dart, so the point goes into the snow, and acts as a wind indicator afterwards). Do not hover, don't go beyond the marker, and try to land just as the snow cloud develops.

Tip: Multibladed aircraft tend to have shorter blades and a more condensed downwash, so the snowball tends to come a little later in your approach. This can make things very white very quickly, right when you really need to see.

Another method, possibly better for beginners, is to come to a high hover (with escapes) to allow the downwash to clear the snow, then lower the machine slowly and smoothly, so you need to be wary if you're heavy (note that the snow cloud will rise, so don't hover too low.. Look for visual references inside the rotor disc.

Still another is a really shallow approach on the edge of translational lift, to slowly move forward onto the site with your skids on the ground (only recommended where the snow cover is light and dry).

Tip: Upon touchdown, applying forward cyclic as you seat the machine in forces the front in lower, to keep the tail rotor as high and clear as possible.

Once landed, bounce the skids a little to see if there's a crust, although you should be careful with the Astar as too much downward force on the blades could cause the head to crack, aside from pushing the belly panel up into the controls (watch out for the swinging hook, too, which could cause loss of fuel). The belly landing light is a great snow scoop which could affect the controls as well.

Keep your RPM to flying levels until you're sure you're on firm ground. Always keep the helicopter light on the skids until passengers are clear, regardless of the surface, and don't let anyone out till you're happy:

"I have also had the experience in a longranger of not being able to land. I proceeded slowly into the snow with a good reference. As I settled I added slight forward cyclic to keep the snow clear of the tail. I had to stay in a hover because the snow reached the Pitot Tube and I was still sinking. When my three passengers left the a/c the first guy out was without snow shoes and he just

about disappeared. The second two learned from the first and even with the shoe's sank to their waists. Every place seems to have different consistency of snow depending on wind/temperature/ shade so I tend to go light into a hole for the first time. That way I don't have to worry about settling into something or not being able to hover above a cloud of snow. "

More tips:

"It is also better to use more than one object (as a reference - author), as if you lose one, you may get a second chance if you have another reference to switch to. Judging your speed can sometimes be tricky in flat light, I've seen my point of reference vanish as I passed by it more than once. Having that backup will save your bacon.

Another trick an old timer taught me was to develop a standard approach. No matter the conditions the last phase of an approach to landing should be exactly the same. The theory behind this is that if things go terribly wrong in the last few feet of your approach, nothing changes. If you're coming in with the perfect speed with the disk loaded and all is set up just right, nothing should change from any other landing. If you know you're set up, everything is just right then you loose all reference with a couple of feet to go, if you do everything exactly the same as every other landing, you should AND I STRESS SHOULD arrive at your spot with the spinning side up."

When you commit yourself, however, you will need to check the firmness of the surface, which is not usually a problem at a camp or something, as the ground crew will have done this for you. The danger lies when you're going to an unchecked site for the first time. Touch down lightly without delay, treating it as a sloping ground landing, as the vibration of the helicopter itself can cause ice to crack. Any form of load spreading is a good idea if you can take advantage of it, like landing on a log pad, although your landing gear will largely determine what you can use. As an example, a fully loaded JetRanger on floats weighs 133 lbs per sq. foot, whilst one on skids and bear paws is nearer 400. Whatever you choose, it needs to be twice as thick if you intend staying overnight.

The colour of ice can be a good clue as to its suitability. White or blue is the thickest, and therefore safest, whereas black ice may have running water underneath and will be quite thin (for this reason, avoid inflows or outflows of streams or rivers). Granular, dirty looking ice is melting. Large puddles or sheets of water are also a dead giveaway. However, ice is never really safe - it doesn't matter how many heavy water trucks the customer may have parked there that week, helicopters vibrate a lot more than trucks do or, more particularly, in a different way, and the hole punched through for the hosepipe has already weakened

the structure (any vibration increases the safe thickness required). Also, however thick they tell you the ice is, you have no way of knowing whether it is actually supported by water underneath (the level may have dropped), or whether any running water has eroded the under surface. Neither do you know whether any snow on top has shielded the ice from the cold, or whether the Sun's rays have reacted with bare ice to act as a lens and create temperatures dangerously close to a thaw, during which ice several feet thick can often become composed of long vertical needles, known as candle ice (read those old Hudson Bay survival manuals), although this only tends to happen in the Spring. Always try and land somewhere else first. Watch out on icebergs - they like to split apart, especially between you and your passengers.

1 inch of fresh water ice at -5°C will support 50 lbs of helicopter per square foot of undercarriage surface area. A 206 with skids and bearpaws has a surface contact area of about 8 square feet, so the ice should be about 8 inches thick to support 3200 lbs. At temperatures above freezing point, the ice should be considered unreliable, and salt water needs to be twice as thick. For variations of 500 lbs, add or subtract 1 inch. If you are planning to park for a few hours, increase all thicknesses by at least 50%, but overnight, the surface should get colder. On the right is a rough guide:

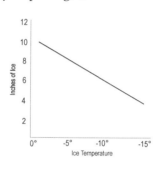

Landings should be smooth and the reduction of collective and power slow, so the transfer of weight to the ice is gradual. This helps you get away quicker if cracks appear or the surface otherwise becomes unreliable. Land as close to a shoreline as possible. If not, the widest part of a river has the slowest current underneath with the thickest ice.

Shut down carefully on an icy surface, anticipating ground spin (especially in a Gazelle). After final shutdown, fill any fuel tanks to prevent condensation, always being aware of your next payload. Remove batteries if temperatures are forecast to be below -10°C. Try to park the machine facing the sunrise, so the Sun's warmth can help with de-icing the windscreen.

Also see that the battery is fully charged before departing from base, and consider taking a spare, together with an external start cable - if the ship's battery runs down completely, the plates will become sulphated and won't actually hold a charge so, even if you start from an external source, you won't be able to start again without one if you shut down. A good ploy is to use the external battery for operating electric fuel pumps, etc. Check that heaters, blowers, etc. work and that snow deflectors are fitted. Also, note whether tie downs and covers are serviceable.

TECHIE STUFF

This chapter deals with some of the more technical aspects of your activities - definitely *not* stuff you learn in flight school! It covers subjects that are frequently required on check rides, and may need revision from time to time.

TWINS

Flying multi-engined helicopters requires a different philosophy in many ways, such as not dumping the collective every time an emergency happens, and their complexity, although there is no real change in flying characteristics as there would be if an engine fails in an aeroplane, aside from a reduction in power available (and you still have to make sure you don't switch off the wrong one!) When multi-crew, emergencies are generally discussed before any action is taken, time permitting (see also *Briefings*, below).

The safest machine is one that can fly on one engine and has one spare, so that both engines, when running, are only working at half speed, which is uneconomical - better fuel economy in the cruise is obtained at the expense of having less powerful engines that take you more gently to the scene of the crash. So, as with most things, there is always a compromise involved.

You must observe the H/V curve (if there is one) if an engine fails. The Bell 212 flight manual states that this is *critical* during takeoff, landing and other operations near the surface.

Note: Being inside the weight limits is especially important with larger helicopters - where a Bell 206 will continue pumping out fuel, for example, when severely overtorqued, something like a Bell 212 will limit the torque available (because it is trying to protect the gearbox) and make the situation worse. In other words, you are less able to get away with a lack of technique!

PERFORMANCE

The regulations require your aircraft to have adequate performance for any proposed flight, meaning its ability (in terms of engine power, at least) to get off the ground in the first place, to maintain certain rates of climb against distance, and land, so you can avoid hard objects (obstacles), particularly when you can't see them.

Many accidents are performance-related, particularly those that happen during taking off and landing. Since you are trying to get a large, heavy object into or out of a relatively small place at some speed, the whole point of performance calculations is to ensure that the space *required* is not more than the space *available*, taking due account of an engine failure right when you don't want it. The idea is *safety*, by keeping the aircraft mass within limits during *all phases of a flight*, as the lighter it is, the better it can fly when less power is available. That is, you don't just plan for the takeoff - you check out the whole route and the part that is most limiting dictates your takeoff weight.

The space required depends on how heavy the helicopter is, and various factors will be working against you, such as runway length, wind, temperature and pressure altitude, which are described below and included in performance charts* in some way. The helicopter (and its power) is not the only consideration - the size of the helipad is important as well because you may have to get back into it.

*A new aircraft comes with a mass of documentation in the shape of the Flight Manual, which is required for certification and which forms part of the Certificate Of Airworthiness. In the Flight Manual will be a series of graphs (in the *Performance* section) showing you how well it will perform under various conditions. For example, you might want to find out how heavy you can be going into a clearing halfway up a mountain. The charts will tell you.

Note: The charts in flight manuals tend to be optimistic, and they are based on new machines and skilled pilots in the first place so, although the graphs will give you a maximum weight for the conditions, you would be wise to give yourself a margin, as the maximum weight is a *limit* and not a *target*. What you can do on one day under a given set of circumstances may well be impossible another time, because particular machines may be better or worse due to the age of the airframe and engines, the maintenance, or crew skill and experience, without the engines being adjusted for several seconds after the initial failure.

As an example, many machines, such as the Bell 212, will not lift the weight stated in the chart. At HOGE-10%, it barely hovers OGE. On the other hand, a B2 will lift way more than Chart HOGE in certain conditions.

It is your responsibility to decide whether or not a safe takeoff (and landing) can be made under the prevailing conditions, although the operator must also ensure that the performance claimed can be achieved. The manufacturer only supplies data in the form of tables and graphs for the relevant Category, which must be matched with the operating environment and declared distances from the site operator (or a survey).

The takeoff and landing phases are the most critical, demanding the highest skills from crews and placing the most strain on the aircraft. It used to be that having enough engines to lift a load was all that mattered and no priority was given to reserves of power. Now it's different, and you must be able to keep your machine a certain distance away from obstacles and be able to either fly away or land without damage to people or property (and the machine) if an engine fails.

For performance purposes, helicopters operate in Classes, 1, 2 or 3, depending on their Maximum All-Up Weight and the number of passengers carried.

They are also certified in Airworthiness Categories, which only dictate the basic ability to fly with an engine out and stand up to stuff like forced landings *when* an engine fails. In other words, the terms *Category* A and *Category* B (as opposed to *Class* 1 or 2) are minimum standards to be achieved by the manufacturer. In the fixed wing world, this would be known as *gross*, or *unfactored* performance.

The Performance Class is an extra layer of operational paperwork that determines the outcome *after* an engine fails and how far away from obstacles you must stay, especially during takeoff and landing. The equivalent of *net* performance, or gross, that is factored in some way.

Note: People often refer to *Category A* Performance when they mean *Class 1* Performance, which is a hangover from fixed wing, which uses Performance Groups. **Remember:**

The machine *operates* in a Performance Class, but is *certificated* in a Category.

More modern helicopters can hover with One Engine Inoperative Maximum Takeoff Mass when Out of Ground Effect (or OGE OEI at MTOM - essential acronyms are explained overleaf), which removes the need to re-land if an engine fails early in the takeoff or late in the landing, so the link between Category A procedures and Performance Class is less, or even broken, although the Cat A procedure is still at the heart of what we do.

Definitions & Abbreviations

Life would not be complete without acronyms. Here is a selection from EASA and ICAO:

- **Absolute Ceiling**. The altitude at which the helicopter is not able to climb at all.

- **Adequate Margin**. Taken to be 35 feet.

- **AEO**. All Engines Operating.

- **Clearway**. An area at the end part of the Takeoff Distance Available (TODA) that is unsuitable to run on, but clear of obstacles, so you can fly over it. Although it is not ground-based, you can include it in your calculations, because you should be staggering into the air towards your screen height just before you reach it.

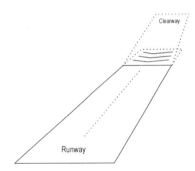

- **Climb Gradient**. The ratio, in the same units, and expressed as a percentage of, *Height Gained* divided by *Horizontal Air Distance Travelled*, in still air. When a headwind is involved, it is *ground distance* travelled.

- **Committal Point**. Most approaches reduce in speed as well as height. By the time you get to, say, 200 feet, there is not enough airspeed to fly away if an engine fails, which is when the decision is made to land on the deck, whatever happens. That is the committal point.

- **Congested Area**. In relation to a city, town or settlement, an area substantially used for residential, commercial or recreational purposes (which may include a golf course - see also hostile and non-hostile environments).

- **Critical Power Unit**. The one whose failure has the worst effect on the situation. Not really a factor with helicopters.

- **CPUI**. Critical Power Unit Inoperative.

- **Distance DR**. The horizontal distance the helicopter has travelled from the end of the TODA (or the *available runway length*), used for adding a divergent quality to obstacle clearances (for example, 0.01 of DR is added when IFR).

- **Defined Point After Takeoff** (DPATO). In Class 2, the point during takeoff and initial climb, before which the helicopter's ability to continue the flight safely, with the critical power unit inoperative, is not assured and a forced landing may be required.

- **Defined Point Before Landing** (DPBL). In Class 2, the point during approach and landing, after which the helicopter's ability to continue the flight safely, with the critical power unit inoperative, is not assured and a forced landing may be required.

- **D Value**. The largest overall dimension of the helicopter when rotors are turning, normally measured from the most forward position of the main rotor tip path plane to the most rearward position of the tail rotor tip path (or most rearward part of the fuselage for Fenestron or NOTAR).

- **Drift Down**. How much height you lose between the time an engine fails and when you can fly away safely under control. The flight path will be mostly vertical in the early stages because your remaining engine(s) will have little power, but as the air density improves, it will become more horizontal. This is an important point if you are over high ground, particularly mountains - you will need to calculate a radius of action to cover this.

  If the machine is likely to level out below MSA, you may jettison only enough fuel for a safe letdown (i.e. to the destination with the required reserves), and not below 1000 feet above terrain.

- **FATO**. Final Approach & Takeoff Area.

- **Helideck**. A heliport on a floating or fixed offshore structure. An elevated heliport is at least 3 m above the surrounding surface.

- **HIGE**. Hover In Ground Effect.

- **HOGE**. Hover Out of Ground Effect.

- **Hostile Environment**. Where a safe forced landing cannot be made because either:

  - the surface is inadequate

  - the occupants of a helicopter cannot be adequately protected from the elements

  - SAR response or capability is not consistent with anticipated exposure

  - there is an unacceptable risk of endangering people or property on the ground

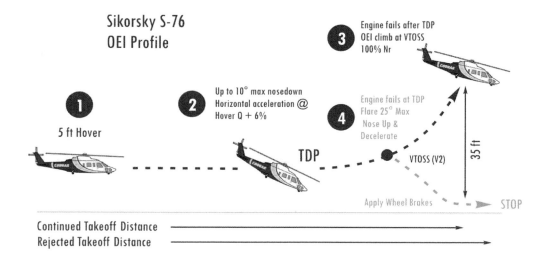

The open sea North of 45°N and South of 45°S, as designated by the Authority of the State concerned, and congested areas without adequate safe forced landing areas are hostile environments.

Congested areas without safe forced landing areas are hostile areas, as are forests and open seas, and mountains, except that mountains are often ignored because there are hardly any helicopters that can maintain level flight over a high range.

• **Landing Distance Available** (LDA). The length of the final approach and takeoff area, plus any more declared available and suitable for helicopters to complete the landing from a defined height.

• **Landing Distance Required** (LDRH). The horizontal distance to land and stop completely* from **15 m** (**50 ft**) above the landing surface (JAR OPS 3.480/CS29.81) (*around 3 kts on water).

• **Landing Decision Point** (LDP). The point from which, with a power unit failure being recognised, the landing may be safely continued or a balked landing initiated. Also the point at which a successful engine-out missed approach can no longer be carried out, and a landing must be made.

• **Large rotorcraft** weigh more than small rotorcraft - up to 20,000 lbs or more

• **MAPSC**. Max Approved Passenger Seating Configuration

• **Non-Hostile Environment**. Where:
  • a safe forced landing can be accomplished
  • occupants can be protected from elements
  • SAR response or capability is consistent with anticipated exposure

• **OEI**. One Engine Inoperative.

• **Precautionary Forced Landing**. An inevitable landing or a ditching in which it can be reasonably hoped that no injuries will be suffered by the helicopter's occupants or by people on the surface.

• **R**. Rotor Radius.

• **PRejected Take Off Distance Required** (RTODRH). This is the distance required for an aborted takeoff - the horizontal distance required from the start of the takeoff run to where the helicopter comes to a full stop after a power unit fails, and rejection of the takeoff at the TDP.

• **Rejected Takeoff Distance Available** (RTODAH). The length of the final approach and takeoff area declared available and suitable for Performance Class 1 helicopters to complete a rejected takeoff. In the fixed wing world, it is called the *Accelerate-Stop Distance Available* (ASDA), or *Emergency Distance Available*, being the length of the runway plus any stopway.

icture: Sikorsky S-76 AEO Landing Profile

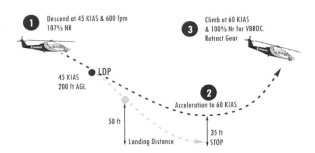

• **Safe Forced Landing**. An unavoidable landing or ditching with a *reasonable expectancy* of *no injuries* to persons *in the aircraft* or *on the surface*. This is an ICAO concept that has been adopted in JAR OPS.

• **Service ceiling.** The altitude at which you can achieve 100 fpm rate of climb.

• **Small helicopters** weigh up to 3175 kg (7 000 lbs)

• **Takeoff Decision Point** (TDP). The first point from which, a power unit failure having been recognised, a continued takeoff is assured, and the last point from which you can still fit into the reject area if an engine fails. It is usually designated as a combination of speed and height, but may only be determined by airspeed, or V_1, which itself depends on pressure height, temperature, wind component, runway available and gross weight.

TDP and CDP (Critical DP) are basically the same thing - the initials DP (Decision Point) are the essential part. Either one is really only critical when operating from a runway which is equal to the accelerate/stop distance for the weight of the aircraft in nil wind for a given temperature, such as small heliports, not fixed wing airfields.

• **Takeoff Distance Available** (TODAH). The length of the final approach and takeoff area, plus the clearway (if provided), declared available and suitable for helicopters to complete the takeoff. This is sometimes also called the Continued Take

Off Distance, which is the total from the start of the takeoff run to where the helicopter would have reached 35 feet above the runway at V_{TOSS}. In fixed wing terms, it is the length of the runway plus stopway and clearway (see *Runway Length*, later).

- **Takeoff Distance Required** (TODRH). The horizontal distance from the start of takeoff to the point at which V_{TOSS}, a selected height (35 ft)*, and a positive climb gradient are achieved, after failure of the critical power unit is recognised before TDP, and with remaining power unit(s) within their limits.

  *The selected height is determined with flight manual data, and is at least the highest of 10.7 m or *35 ft above the takeoff surface*, or a level defined by the highest obstacle in the TODR. For elevated decks, there is no 35 ft requirement.

  This may be disregarded if the helicopter can (after TDP) clear any obstacles* between the end of the TODAH and where it becomes established in a climb at V_{TOSS} vertically by at least 35 ft (10.7 m).

  *Within 30m or 1.5 times the maximum dimension of the helicopter, whichever is greater.

- The **Takeoff Flight Path** extends from the start of the takeoff run until 1000 feet above the surface, to include periods of acceleration and climb, **remaining clear of the H/V envelope***. The helicopter must not approach the surface closer than 15 ft (when the TDP is more than 15 ft), and there may be descent below the level of an elevated deck if everything is avoided by 15 ft.

  *In theory, a Cat A procedure should be outside the H/V curve, but, for some aircraft, the curve can be modified to cope (check the flight manual). However, the H/V curve does not take account of the dynamics associated with a profile, so profiles that look as if they creep into the H/V curve will have been tested and found to be safe.

- **TLOF**. A load bearing area on which a helicopter may touch down and liftoff.

- **Unaccelerated Flight**. The hover or steady straight and level flight, climb or descent.

V-Speeds

These are significant operating speeds:

Speed	Explanation
V_{LE}	Max gear extended
V_{LO}	Max gear operating
$V_{maxRange}$	Speed for maximum range
V_{maxend}	Speed for maximum endurance
V_{NE}	Never Exceed speed.
V_{NEI}	Never Exceed speed - IFR
V_{MINI}	Minimum IFR speed
V_{NO}	Normal Operations. 10% less than V_{NE}.
V_{stayup}	The ability to continue and accelerate without descending
V_{TOSS}	Takeoff Safety Speed in a **Category A helicopter** (equivalent to V_2 in an aeroplane). It was developed because it is below V_Y, and repeatable, aside from giving a better angle of climb than rate of climb, because you're more interested in clearing ground in a short distance than how fast you're going up. In other words, it gives good angles and rates of climb, but not the best, while being high enough to allow the aerodynamic surfaces to kick in for stability purposes. It is limited at V_Y. On the S-76, it is 110 kts higher than V_1.
V_X	Best angle of climb, or the most height in the shortest distance. It does not apply to helicopters because, although, in a vertical climb, it would be zero, at all other times it would be so variable as to be impossible to calculate and below a reliably indicated airspeed anyway. See V_{TOSS}.
V_Y	Best rate of climb, or the most height in the shortest time, occurring with the greatest difference between power available and power required. Also minimum power airspeed.
V_{YSE}	Best rate of climb, single engine
V_{BROC}	See V_Y.

Note: Best angle of climb speed increases with altitude. Best rate of climb speed stays constant.

Airworthiness Categories

CATEGORY A

This provides for multi-engined helicopters with engine and system isolation, under a critical engine failure concept that allows a controlled landing or continued flight because of redundancy (another engine) or a design process that limits failure in the first place. *Engine isolation* means that one engine failure should not lead to a second, and that fire in an engine compartment can be detected, contained and/or extinguished (for 15 minutes).

In English, if an engine fails after you start to take off, you (or rather, the helicopter) should be light enough to either return to and stop safely on the takeoff area (before TDP), or continue and climb out (after TDP), maintaining minimum rates of climb at certain points. The takeoff area should be large enough to manoeuvre in if you cannot. For example, an elevated site will have less lateral visual clues than a ground level site, so it might need to be larger.

Category A also specifies the data to be put into a flight manual that is needed to calculate the landing (or re-landing) areas if an engine fails, plus OEI obstacle clearance for the Performance Class. More comprehensive flight and navigation equipment is also listed.

Picture: Category A Takeoff Profile. In the fixed wing world, each part of the profile represents a different aircraft configuration (gear or flaps up, etc.)

Category A compliance assumes:

- Normal piloting skills and no extraordinary environmental conditions

- Still air at sea level in ISA conditions

- Relative humidity of 80% in ISA conditions for piston engined helicopters

- Relative humidity of 80% at and below ISA temperature or at and above standard temperature plus 28°C for turbines. Any between those limits must vary linearly

PROFILES

A Category A takeoff or landing profile is a set of manoeuvres that provide continued safe flight or a controlled landing after an engine fails. **It does not take account of obstacles!** Single-engined machines have profiles as well, which should keep them out of the Height/Velocity curve by about 10 feet and 5 knots.

Note: Profiles must be flown accurately! A few knots either side of the target speed will mean that you will not climb and may even descend (not that the climb rate will be startling in the first place!) Profiles can often be flown more accurately by an autopilot.

The usual Category A takeoff profile has four sections, each of which can affect your takeoff weight. It includes a first segment which requires the ability to climb at 100 ft/min at V_{TOSS} (at least) from the end of the TODRH to 200ft, a second segment* which needs a climb performance of 150 ft/min at V_Y at 1000 ft above the takeoff surface (on which a safe forced landing can be made) and a level acceleration from V_{TOSS} up to V_Y (may be combined with the first segment climb).

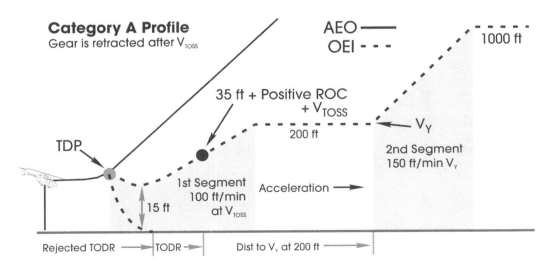

© Phil Croucher, 2016

If this page is a photocopy, it is not authorised!

*The second segment is usually the most limiting because it has the greatest gradient.

There are similar arrangements for landing.

Note: Using a profile does not automatically make you Category A compliant! Category A procedures only provide basic tools to establish reject distances and climbs.

The flight manual may contain tables with variable TDPs, drop downs, reject distances, V_{TOSS} and TODR, so you can tailor your profile to fit the takeoff site.

PHASES OF FLIGHT

Levels of performance must be considered for these phases of flight, with one engine inoperative:

- **Takeoff.** You must achieve rates of climb of 100 ft/min at 60 m (200 ft) and 150 ft/min at 300 m (1000 ft). Up to the Takeoff Decision Point (TDP), you must have the surface in sight, for a rejected takeoff if necessary. Landing gear may not be retracted until V_{TOSS} and a positive rate of climb have been achieved.

- **Takeoff flight path.** You must clear all obstacles by 10.7 m (35 ft) when VFR, + 0.01 DR when IFR

- **Enroute.** When not visual with the surface, you must be able to climb at 50 ft/min at 300 m (1000 ft) (600 m in mountainous areas), above all obstacles within 18.5 km (10 nm) of the intended track. When visual with the surface, the obstacle limit becomes within 900 m of the intended track.

- **Descent.** The OEI flight path must take the helicopter down to 300 m (1 000 ft) above the landing area, clearing all obstacles within 18.5 km (10 nm) of the intended track by 300 m (1 000 ft) or 600 m (2 000 ft) in mountainous areas. When visual with the surface, the obstacle limit becomes within 900 m of the intended track

- **Landing.** You must be able to maintain the same minimum rates of climb as on takeoff, that is 100 ft/min at 60 m (200 ft) and 150 ft/min at 300 m (1000 ft), *at the estimated landing time, above the destination and alternate.*

CATEGORY B

This covers single-engined helicopters, or multi-engined ones that do not fully meet Category A requirements (for example, there may not be enough engine isolation, or manufacturer's data). They are not guaranteed to stay airborne if an engine fails and an unscheduled (i.e. forced) landing is assumed.

The fact that a helicopter has two engines does not necessarily make it Category A compliant. For example, the Bell 212 is Category B under normal circumstances. For vertical takeoffs and landings from heliports (i.e. Cat A), certain conditions must be met:

- An approved copilot kit, dual controls and a suitable altimeter

- Two pilots

- All doors installed and closed, heaters and vents off until CDP (to prevent altimeter errors)

- RRPM 100%, skid landing gear

- Max weight 10,000 lbs suitably adjusted for WAT

- C of G restrictions

- Heliport size 72 x 150 feet

- Max altitude 2500 feet PA

The essential difference between the categories is that Category A has a guaranteed stay-up capability, while Category B has not.

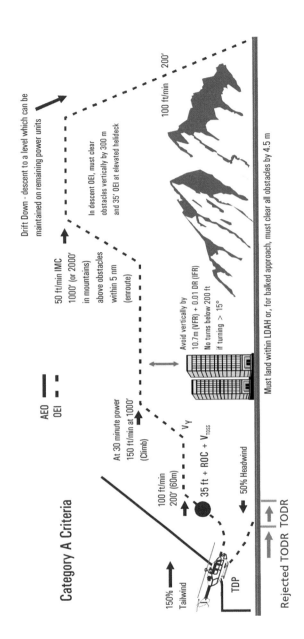

Category A Criteria

Drift Down - descent to a level which can be maintained on remaining power units

50 ft/min IMC
1000' (or 2000' in mountains) above obstacles within 5 nm (enroute)

In descent OEI, must clear obstacles vertically by 300 m and 35' OEI at elevated helideck

100 ft/min 200'

Avoid vertically by 10.7m (VFR) + 0.01 DR (IFR)
No turns below 200 ft if turning > 15°

Must land within LDAH or, for balked approach, must clear all obstacles by 4.5 m

AEO
OEI

At 30 minute power
150 ft/min at 1000' (Climb)

V_Y

100 ft/min
200' (60m)

35 ft + ROC + V_{TOSS}

50% Headwind

150% Tailwind

TDP

Rejected TODR TODR

---

worry about obstacles in the obstacle accountability area, which is what the Performance Class is all about.

- Helicopters with more than 19 passenger seats, or operating to or from heliports in congested hostile environments, *must* operate under Performance Class 1

- Helicopters with 9-19 passenger seats may operate under Class 1 or 2

- Helicopters with 9 or less passenger seats may operate under Class 1, 2 or 3

A single-engined helicopter will automatically come under Class 3.

Class 1

This class applies to helicopters with a MAPSC of greater than 19, or those within congested, hostile environments. It requires no forced landing provisions if the critical power unit fails - the machine can either land within the rejected takeoff distance or continue (safely) to a suitable landing area, depending on when the failure occurs (that is, before or after TDP), clearing all obstacles vertically by 35 feet with an engine out (plus a percentage of DR when IFR*), being able to achieve a rate of climb of 100 ft/minute at 200 ft.

In other words, a twin-engined helicopter must be able, if the critical power unit fails at or before TDP, to discontinue the takeoff and to stop within the rejected takeoff area available, or, if the failure occurs at or past TDP, to continue the takeoff and then climb, clearing all obstacles along the flight path by an "adequate margin", which in ICAO-speak is 35 ft.

In this way, there should be no chance of an accident if an engine fails *at any stage of a flight* (this applies to all classes!)

In addition, the site must be surveyed and be of good enough quality for a reject area. The site must also be under the control of the heliport staff.

Performance Classes

Performance classes concern the conditions under which data in the flight manual is used by ensuring that there is a safe margin between what the helicopter is *able* to achieve and what it is *required* to achieve, because machines vary.

Category A procedures provide non-adjusted profiles - that is, they specify only the climb performance required by the rules. Only when the Take-off Flight Path (which starts where TODR is established) is specified do you

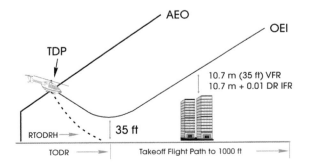

AEO
OEI
TDP

10.7 m (35 ft) VFR
10.7 m + 0.01 DR IFR

RTODRH
35 ft
TODR
Takeoff Flight Path to 1000 ft

You will also need a profile to match the site - that is, the profile must meet PC1 obstacle clearance requirements.

Note: For many helicopters, full Class 1 performance off an elevated site such as an oil rig requires the takeoff weight to be significantly reduced, especially if the wind speed is low, or from the wrong direction, or the temperature is high. *This could be around 50% of normal takeoff weight.* That is, a 10-seat helicopter may have to reduce its passenger load to 5!

Very few helicopters can operate in Class 1 from an offshore deck with more than a few passengers. Being committed at the point of rotation doesn't mean you can actually fly away, it just means that you are committed to continuing the take off to get enough airspeed to at least fly level or, even better, start to climb before you enter the water (i.e. you have no choice). Under most jurisdictions, even Class 2 operations require that a twin engined aircraft must be capable of Class 1 performance at some point, often 1000 feet above the take off elevation.

Important factors include takeoff and landing distances (size of heliports), rates of climb (or descent) against obstacles, whether you are IFR/VFR or HOGE/HIGE.

Only from the takeoff flight path do you need to take account of obstacles.

Backup/Lateral Procedures

These demand special consideration because there are fewer cues to rely on, aside from the sight picture through the front window, and the altimeter.

There are three variations on the Class 1 takeoff theme that comply with Cat A:

CLEAR AREA

As if taking off from a runway, with nothing in the way. It satisfies legal minima as long as you can maintain 100 ft/min at 200 feet with one engine out, and 150 ft/min at 1000 feet. A fixed TDP is normally used because there is nothing to restrict your ground run or flight path.

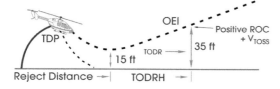

The helicopter is assumed to accelerate down the FATO (runway) outside the H/V curve. If the engine fails after TDP, it may lose height down to a specified height, which is usually 15 ft, if the TDP is higher than 15 ft.

For the Sikorsky S-76, once the bugs have been set for TDP and V_2 on the ASI, establish a 5-ft hover. Increase collective pitch until you reach 6% above that used for the hover and accelerate forwards, keeping the wheels 5-10 feet off the ground until reaching TDP (the exams may use the older CDP).

After passing TDP, rotate nose-up to initiate the climb at V_2. Once clear of obstacles, gradually accelerate to best rate of climb speed (V_Y) and retract the wheels.

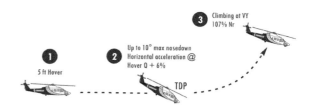

Bleed Air should be Off, and Anti-ice may be On or Off.

RESTRICTED AREA (SHORT FIELD)

With nearby obstacles that must be cleared by 35 feet (the min-dip), and a surface on which a reject can be carried out. This requires a steep or vertical climb before proceeding forward over the obstacles. The TDP can be varied to give an improved clearance (or clear a higher obstacle), but as its height increases, it is more difficult to land back in the reject area:

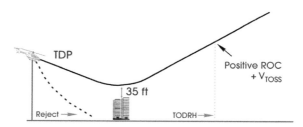

The distance to the obstacle does need to be calculated (as long as it is outside the reject distance) as you will be missing it by 35 feet anyway.

HELIPAD

The reject area is the helipad itself, so you must keep it in sight to re-land on if the engine fails before TDP.

- For *Non-elevated Heliports* (i.e. at ground level), the takeoff to TDP is a rearwards climb, if there are no obstacles behind. If there are (say on an oil rig), TDP will normally be at or below 30 feet:

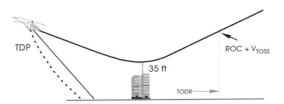

The point where V_{TOSS} meets a positive rate of climb defines the TODRH.

This is an example for the AS 355:

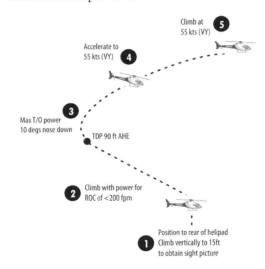

- For *Elevated Heliports* and *Helidecks* (e.g. oil rigs, or anything at least 3 m above the surrounding surface) the same applies, but you have extra space in which to drop down and gather some speed:

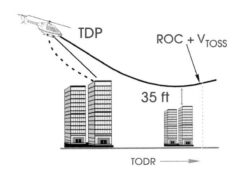

LANDING

This is a PC1/Category A Landing procedure:

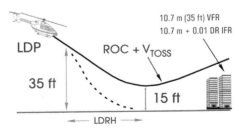

There are two requirements - all obstacles must be cleared by 35 feet in the approach to land, and the helicopter must be able to land and stop in the distance available. The part of the landing between LDP and the touchdown point should be done with the surface in sight. On a balked approach, you must clear all obstacles vertically by 10.7 m (35 ft), and you may not descend below 4.6 m (15 ft) above the surface.

In the Sikorsky S-76, arrange the approach to reach LDP at 200 feet above the touchdown elevation at 45 kts, with 107% N_R and a rate of descent up to 600 fpm. Start deceleration after passing through 50 feet at 45 KIAS. Continue the approach until a running touchdown or a hover is achieved.

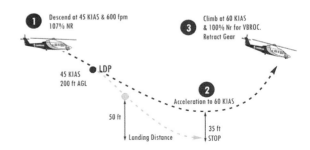

By way of contrast, this is an example for the AS 355:

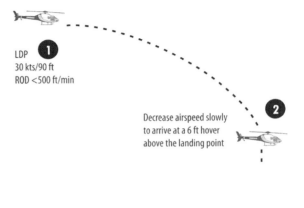

Class 2

Performance Class 1 needs a surveyed area in which all obstacles in the takeoff path have been taken into account. The offshore environment has moving obstacles (boats) and PC1 departures are impossible, hence PC2 enhanced, which is a PC1 takeoff without the requirement for a surveyed takeoff path. The subdivisions of PC 2 are purely about what happens when the engine fails early in the takeoff or late in the landing. The part of the takeoff path in which an engine failure would result in a forced landing must be flown in conditions of weather (600 ft cloud base) and light (800 m visibility) in which a safe landing is possible, so a reject area is not specified.

Here (and for Class 3), there is less clarity as an element of pilot judgment* is required - there is no ability to climb away if an engine fails before DPATO or after DPBL.

*Unfortunately, DPATO is a bit of a woolly concept, because it has to satisfy a number of requirements that do not always come together at the same time. In fact, there are three aspects to a PC 2 takeoff:

- The bit before TDP for the basic Cat A procedure.

- The part after DPATO, where OEI obstacle clearance is established.

- The bit in the middle, during which you need some pre-considered action if an engine fails, so a safe forced landing area need not be calculated, as it might be for the period before TDP.

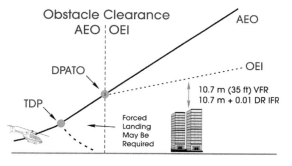

If you have to make two calculations (one to establish the safe forced landing distance, and one to establish DPATO) using two Cat A procedures, the more limiting mass should be used.

So how do you calculate DPATO, which is the starting point for your obstacle clearance in terms of speed and height above the takeoff surface? ICAO Annex 6 has no guidance on how to figure it out. The simplest way is to work out the AEO distance needed to get to 200 feet at V_Y. Alternatively, you could use the distance needed to get

to 50 feet, which is the starting point for landing calculations, if your reject area is large enough. Or you could just use the standard Category A clear area figures, which would be conservative - the idea is to have the V_Y at 200 feet point as close as possible to the takeoff point.

Distances do not have to be calculated when, by using pilot judgment or standard practice, a safe forced landing is possible and obstacles are cleared AEO (takeoff) or OEI (climb), but Annex 8 (Airworthiness) requires that a distance be scheduled for Performance Classes 2 and 3. You cannot go IFR before DPATO because the climb gradient will not have been established. A forced landing is less of a problem on land, but things are a little less cut and dried when offshore, as the reject area is the water. ICAO Annex 6 Class 1 performance requirements are written mainly for land-based operations, so ground effect or reject areas can be discounted.

However, once past DPATO, or up to DPBL, the same OEI performance (and obstacle clearance) as Class 1 is expected to be available. The risk element is minimised because Performance Class 2 helicopters are certificated in Category A (i.e. they at least have two engines).

This is a typical PC 2 takeoff from an elevated deck or FATO in a non-hostile environment,:

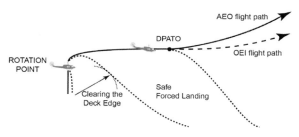

If an engine failure happens before the rotation point, you should be able to put the machine back on to the deck. After rotation and before DPATO, you should be able to perform a safe forced landing. After DPATO, the OEI flight path should allow you to clear obstacles by the required margins.

At normal operating weights this means that, until the helicopter has got some forward airspeed to assist with performance, it is often unable to climb away on one engine if an engine fails, and may hit the edge of the deck. Once it has enough speed, it should be able to comply with Class 1 at normal and economic operating weights (i.e. 150 fpm ROC at 1 000 ft). However, with effective risk management, such as HUMS, this exposure is considered to be acceptable for day to day operations.

EXPOSURE TIME

Exposure is the period during which OEI performance does not guarantee a safe forced landing or continuation of flight, so such operations do not need you to provide for safe forced landings (although you still have to account for obstacles). In the picture below, for a non-hostile environment, the exposure time is between the rotation point and DPATO:

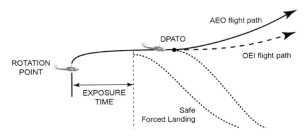

If an engine fails between the end of the exposure time and DATO, a safe forced landing should be possible. After DPATO, the OEI flight path should clear obstacles by the required margins.

This pictures shows a typical takeoff from an elevated deck in a non-congested hostile environment.

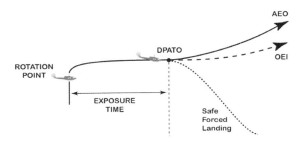

If an engine fails after the exposure time, you can carry out a safe forced landing as, after DPATO, you should be able to clear obstacles by the required margins.

The takeoff should be made dynamically, using between 110-120% of the power used in the hover up to the rotation point to keep things under control (i.e. not too fast, not too slow). You need to continue upwards and away from the deck if an engine fails, without ballooning, but also not hitting the edge of the deck if you lose height.

The *maximum permitted exposure time* is a statistically derived figure, during which the probability of an engine failure can be discounted or, put another way, during which you hope an engine doesn't fail*. In fact, the probability is "extremely remote". The maximum permitted time should naturally not be more than the actual exposure time.

*Data from the OGP Aviation Safety Subcommittee shows that engine-caused fatal accidents are very rare - in multi-engined helicopters, engine failure is no longer a serious safety concern.

On an elevated deck, **single-engined** exposure time starts from the nose-down cyclic input until the deck edge is cleared. This is typically about 4 seconds. A safe forced landing is assumed once the deck is cleared (which really means that no provisions have been made for one - you're on your own).

For **multi-engined** helicopters, it starts from when a safe forced landing on the helideck is no longer possible and stops when an engine failure would not take you below the helideck level. In other words, there is no drop down*, although one of 50 feet could reduce the exposure by 1 second from the typical 6 seconds. Exposure on landing begins when you can no longer fly away and ends when you can actually land on the deck (at the committal point).

*This is dropdown:

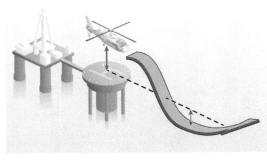

The takeoff mass for operations with exposure is taken from the more limiting of:

- the climb performance of 150 ft/min at 1 000 feet above the takeoff point.

- obstacle clearance.

- AEO HOGE performance, to ensure the possibility of acceleration when using takeoff techniques that are anywhere near vertical. There is also a power reserve for elevated platforms.

The absolute limit for exposure is 200 feet.

LANDING

This is a typical landing profile for an elevated helideck in a non-hostile environment (no exposure):

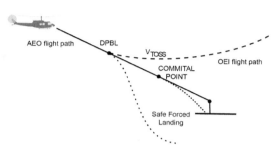

The DPBL is a window in terms of airspeed, rate of descent and height above the landing surface. If an engine fails before DPBL, you can land or carry out a balked landing, for which you must clear all obstacles vertically by 35 feet.

If the failure happens between the DPBL and committal point, you should be able to make a safe forced landing on the surface. After the committal point, land on the deck.

If you add exposure to the mix:

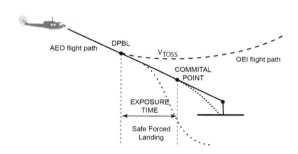

You can do a balked landing before DPBL or a safe forced landing after it, or the deck if you make it to the end of the exposure time.

Finally, for a non-congested hostile environment:

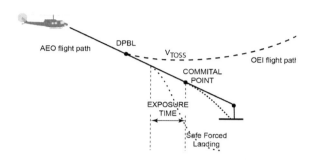

You should be able to continue the flight without hitting anything if an engine fails before the start of exposure. If it happens after exposure you should be able to make a safe forced landing on the deck.

CLASS 2E

Enhanced PC 2 is an attempt to make Class 2 operations attain those in Class 1. So why not use Class 1? The answer is that there is likely something missing - for example, performance data may not be available (the Bell 212 flight manual does not have drop down figures), or the helipad may be too small or have obstacles that stop you using a Category A procedure. For 2e, the helicopter mass should be that needed for an AEO OGE hover. There is no sea state limit. The aim is to reduce risk in hostile areas and complements PC2-with-exposure, the ultimate aim being to reduce exposure to nil.

PC2e demands calculations of mass at takeoff and landing to take into account the following:

- Takeoff and landing profiles and/or procedures

- Deck edge miss distance

- Dropdown height available to acquire OEI climb airspeed after an engine failure

- Weather conditions at the time

OBSTACLES
As for Performance Class 1.

BENEFITS OF CLASS 2

- You can use Category A profiles and distances when the surface is not good enough to be a reject area but is good enough for a safe forced landing

- You can operate when the safe forced landing distance available is outside the heliport boundary (and all you need is to legally hop over the fence)

- You can penetrate the H/V curve for short periods during takeoff and landing

Class 3

Helicopters operated in Performance Class 3 can be certificated as Category A or B, except single-engined machines which are automatically Category B (FAR 27).

xposure for Performance Class 3 is limited to the takeoff and landing phases, in non-hostile environments, when approved. This means that, once you have cleared any obstacles by an adequate margin, and have reached V_Y or 200 ft, a safe forced landing must be possible. That is, you can operate into or out of a clearing, but not one in the middle of a forest (approvals are not intended for continuous operations anyway). On an elevated helideck, the safe forced landing must be possible after reaching the knee of the H/V diagram.

Procedures

Which (performance) group you belong to depends on:

- Certification

- Max All-Up Weight

- The number of passengers carried

However, it may be more acceptable commercially to operate at a lesser performance level if it enables you to take more payload, and make more money - all you might need is longer takeoff runs or less obstacles. In other words (just to reinforce the point), the performance conditions under which you operate determine how heavy your aircraft can be and, as a result, your payload. Over a whole trip, the weight could be dictated by:

- Maximum weight

- WAT limits (see below)

- Space available

- Obstacles

- The route

- Hovering OGE (you may have to accelerate OGE off a rig or small site)

There are some principles with performance calculations that remain constant:

- You can only *plan* to use 50% of any reported headwind component, and *must* use 150% of any tailwind (reported), unless the wind measuring equipment is "approved", in which case you can use 80% of the headwind component

- You must account for weight, altitude and temperature (WAT) for the destination as well. Even this may restrict your takeoff weight

- Distances *required* must not be more than distances *available*

- The helicopter mass must not be more than that in the flight manual for the conditions

INDEX

A

ADF 8-13
Aerial Application 3-1
aerial clipping 3-26
Aerial Filming and Photograph 3-21
Aerial Harvesting 3-26
Aerial Ignition 3-8
Aerial Survey 3-22
AFFF 3-12
air conditioning 4-1
Air Testing 3-22
Air Transport Licence 3-15
Ambulance Taxi Transport 3-10
American Meteorological Society 8-18
anabatic 5-10
AOC 3-15
Application Forms 1-7
Arctic 8-10
Argos 4-13
AS 350 5-31
ASI 3-16
AStar 3-6, 3-22
ATT 3-10
autorotations 8-15
Avalanche Control 3-25

B

Back burning 3-8
baggage 8-10
baggage compartment 8-11
Bambi Bucket 3-9
Bambi Fire Bucket 3-6
BCF 3-12, 3-15
bearpaws 5-11
Bergwind 5-9
Billy Pugh 6-4
bird dog 3-5
Blankets 8-11
Block Fuel 8-5
Bodies and Remains 3-19
Bowden cable 4-4
Boyles Law 5-2
BPC12 3-12
BPC9 3-12
brace position 8-14
Breaklink 6-4
Brownout 8-19
Bucketing 3-7

C

C of A 3-21
C of G 3-20, 3-21, 3-26, 4-12
Cable cutters 6-4
camera mount 3-21
cap cloud 5-14
cargo door 6-1
CARs 3-20, 4-2
catpaws 5-11
Cedar salvage 4-1
Chinook 5-14
chokers 4-1, 4-4
CL 415 3-5
Class A load 4-2
Class D External Load 6-1
Clear Air Turbulence 5-12
clearing 5-31
Clearway 9-2
clevis 4-4
cloudbase 3-18
CO2 3-15
Coast Crawl 6-5
cold soak 8-16
Collective Bounce 4-16
commentary 6-6
Commercial Air Transport 3-16
con 6-6
confined area 5-31
Congested Area 9-3
congested areas 3-16
contact gloves 8-15
Countour Search 6-5
crew 6-2
crop spraying 3-1
crowning fire 4-17

D

D Value 9-3
Decca Navigator 3-22
De-icing 8-12
deicing fluid 8-16
Density Altitude 5-2
Deplaning 3-20
Designated Mountain Areas 5-1
dinghy 8-8
Distance DR 9-3
Ditching 8-7
Double Lift Harness 6-3

downdraughts 5-8
downwash 4-22
DPATO 9-3, 9-11
DPBL 9-3, 9-11
Drift Down 9-3
drip torch 3-9
Droppable Containers & Packages 8-11
Drums 8-5
Dry Chemical 3-12
dynamic rollover 3-26
Dynanav 3-23

E

earthing 6-1
ELT 8-8
Emergency Equipment 8-11
Emergency Response Plan 3-24
Emplaning 3-20
employer 1-2
estuaries 8-1
evaporation 3-4
Expanding Square 6-4
exposure time 9-12
Extra Fuel 8-5
Eye-Level Pass 5-21

F

FARs 4-2
FATO 3-13
Feeder Sites 8-2
filters 8-2
Final Approach and Takeoff Area 3-13
final reserve fuel 8-4
First Aid kit 3-12
Flat light 8-19
flight computer 5-2, 5-3
flight idle 8-18
Flight Manual 4-3
Flight Times
    Recording 8-13
flight watch frequency 3-5
FLIR 3-6
flying club 1-1
foam 3-12
focal length 3-22
Fohn 5-14
forests 3-1
formation 3-22

Formation Flying 3-19
frost, ice and snow 8-16
Fuel 8-2
Fuel Checking 8-6
fuel consumption 4-12
Fuel management 8-4
Fuel reserves 6-3

G
GPS 3-22, 6-15, 8-10
Grand Prix 3-15
Ground Effect 9-3
Ground-Air Visual Signals 8-11
Guideline 6-4
Gusts 5-10

H
handlers 4-2
headwind component 9-14
Height Corrections 6-7
Helicopter Emergency Medical
Service 3-9
helideck 7-5
HEMS 3-9
Herbicides 3-3
HF 4-3
High Intensity Strobe Lights 3-17, 6-1
HISLs 3-17, 6-1
Hoist control 6-4
Hoist Harness 6-3
Hoist Operator 6-6
hoist operator 6-3
Hoisting area 6-2
Hoisting checks 6-15
Homing 6-4
Horse Collar 6-3
Hostile Environment 9-3
hot spots 3-6
hotspots 3-8
Hover Emplaning & Deplaning 3-5
Hover Emplaning and Deplaning 3-20
humidity 5-4
Hypothermia 8-10
hypothermia 8-10

I
icebergs 8-21
Icing 8-12
ICT 3-10
IFR 8-14
Illusions
mountains 5-4
IMC 8-14
Immersion Suits 8-9
In-flight replanning 8-5

Initial Attack, 3-6
insecticides 3-3
insurance 3-17
insurance companies 1-2
Intensive Care Transport 3-10
Interview 1-7

J
Jet A 8-3
Jet B 8-3
jettison 9-3
Journey Logs 8-13
JPT 8-18

K
katabatic 5-9
katabatic wind 5-14
Kevlar 4-18
Kodiak 3-23

L
land breeze 5-10
landing ban 7-2
Landing Decision Point 9-4
Landing Distance Required 9-4
landing pad 8-2
Landing Sites 8-1, 8-20
mountains 5-17
landing sites 8-14
lanyards 4-1
Last Known Position 6-4
last known position 6-5
LD50 value 3-3
LDA 9-4
LDP 9-4
LDR 9-4
Levels of SAR Service 6-2
Lifejackets 8-9
lifejackets 8-14
liferaft 8-8
Line And Hover Corrections 6-6
Line Patrol 3-16, 3-18
Load Behaviour 4-14
Load swing 4-15
loaders 4-14
Log Pads and Platforms 5-18
Logging 4-1
Longlining 4-17
longlining 3-26, 4-20
LongRanger 3-6

M
manual release 4-4
marshaller 4-14
Marshallers 8-18

marshallers 3-12
Medical Supplies 8-11
Minimum Reserve Fuel 8-5
Minimum Sling Specifications 4-10
mirror 4-4
Mountain wave 8-19
mountain wave 5-12
mountainous areas 6-5
mountains 4-22
Movies 3-21
Murphy's Law 8-11

N
navigator 3-26
Night Flying 8-14
No-hover landings 5-19

O
OAT 8-12
obstacle 8-1
Open water hoisting 6-2
Operational Area 3-15
ouchdown and Liftoff Area 3-15
overdue aircraft 8-14

P
Parachute Dropping 3-20
Particle size 3-2
Performance
Helicopter Profiles 9-6
peripheral vision 3-26
Persons under the influence 3-18
photography 3-21
Pipeline Survey 3-18
planking 4-15
Pleasure Flying 3-11, 3-15, 8-2
Police 3-18
Police AOC 3-18
Police Dogs 3-18
Police Operations 3-5
Power Line Cleaning & Maintenance 3-17
PPL 1-3
Precipitation 8-19
Pre-Hoist Check 6-3
Pressure altitude 5-2
Prisoners 3-18
Protective Clothing 8-11
protective gloves 4-3

Q
QFE 5-21
Quads 4-13

R

radar 6-15
radios 8-13
Rafts 8-9
rappelling 3-6
Rate One turn 6-15
refuelling 3-13
Rejected Take Off Distance Required 9-4
Rejected Takeoff Distance Available 9-4
relative humidity 5-4
Release tools 3-12
Remote Areas 8-10
remote installation 7-2
Rescue Device 6-16
Rescue Equipment
     Pleasure Flying 3-11
Rescue Net 6-4
Rescue Strop 6-3
Reserve Fuel 8-5
Resume 1-5
Retardant 3-5
ridges 5-7
Rockies 5-1

S

Safe Forced Landing 9-4
Safety Area 3-15
safety glasses 4-3
SAR 8-9
SARBE 8-8
Scaffolding 4-15
Schermuly flares 8-15
Sea Breeze 5-10
Sea Movement 8-8
Search Patterns 6-4
Seat belts 3-20
SEAWATER 8-9
Sector Search 6-5
Seeding 3-4
Selective logging 4-1
service ceiling 9-4
shallow approach 5-22
shock loading 4-7
Sling Equipment 4-4
Slung Loads
     Preparation 4-12
Spacecam 3-22
Special VFR 3-6
Specific Fuel Consumption 8-4
Spectra 4-18
Speed Corrections 6-6
Spray Drift 3-4
staging area 3-24
Static 8-17

Static Electricity 4-3
static electricity 6-1
static lines 3-20
Stokes Litter 6-4
strong wind conditions 5-21
Sun Tables 8-20
survival equipment 8-11
Survival kits 8-10
Survivors 8-11

T

Tag lines 4-13
Takeoff Decision Point 9-4
Takeoff Distance Available 9-4
Takeoff Distance Required 9-5
Takeoff Flight Path. 9-5
tape recorders 8-13
Taxiways 3-15
TDP 9-4
Teflon 8-6
TLOF 3-15, 9-5
TODA 9-3, 9-4
TODR 9-5
towing 4-2
Township System 3-24
Track Crawl 6-5
training background 1-3
Trip Fuel 8-5
turbine fuel 8-6
Twin Otter 8-3
Twins 9-1
Twinstar 3-22

U

US Army 8-1
Utility Hoist 6-3

V

valleys 5-24
Vapour drift 3-4
Vertical Reference 4-17
Vertical reference 3-9
Vessel 6-15
VFR 8-14
VHF 4-3
Visual Signals 8-11
VMC 8-1
VNE 4-3, 4-12, 4-15, 5-8

W

Water bombers 3-5
water droplets 8-4
Water Fog 8-19
water paste 8-4
Water-finding paste 8-6

Weapons and Munitions 3-19
Weather minima 6-2
Whiteout 8-18
Wildlife Capture 3-26
Winch Cable 6-3
winch cable 6-1
wind reversals 5-8
Winds
     mountains 5-8
Winter Operations 8-15
Wires 3-17
wirestrikes 3-17
wire-stringing 4-2
witness marks 4-4

Y

Y Strap 6-4

Z

Zeiss trilens 3-22